The New Farmer's Almanac

GRAND LAND PLAN

D1637087

VOL. V

GREENHORNS

The New Farmer's Almanac, Volume V
Grand Land Plan

Published by the Greenhorns
greenhorns.org

GREENHORNS

Copyright 2021 by Greenhorns
To reprint original work, request permission
of the contributor.
Printed in the USA by McNaughton & Gunn.
ISBN 978-0-9863205-3-8

The publication of this book has been made
possible in part by support from Furthermore: a
program of the J. M. Kaplan Fund.

Furthermore:
a program of the J. M. Kaplan Fund

Editor in Chief
Severine von Tscharner Fleming

Lead Editor
Briana Olson

Visual Editors and Designers
Tatiana Gómez Gaggero and José Menéndez

Copy Editors and Proofreaders
Alli Maloney
Sue Duvall
Jamie Hunyor

Cover Design
Katie Eberle and Tatiana Gómez Gaggero

Typefaces Used
Tiempos Text, Tiempos Headline, and National
by Klim Type Foundry

Paper Used
55# Natural Offset Antique, FSC Certified,
360 PPI

THANK YOU
Thank you to our readers for all your kind
feedback, which keeps us publishing, and for
pushing your talented friends to write essays
and submit stories and artworks. To all our
writers, editors, and contributors—thank you
for sharing your love and your public thinking.
To Filson, Furthermore, and the Flora
Foundation—thank you for making this project
possible. To Rick and Megan Prelinger,
for instigating the first Almanac. For new and
longstanding support, thank you to Carolyn
Murphy, Alex Carleton, and Alessandro
D'Ansembourg. Thank you to Briana Olson, our
fearless editor without whose meta-research,
massaging, reminding, demanding, and
diligent follow-ups there would be no book at
all. And thank you to the tireless pursuit of
quality held by our evolving Almanac team,
including Tatiana Gómez Gaggero, José
Menéndez, Katie Eberle, Nina Pick, Charlie
Macquarie, Audrey Berman, Hallie Chen,
Tess Diamond, Francesca Capone, and Nicole
Lavelle. You guys rock!
    —SVTF

Thanks to our contributors, for your generosity
in labor and spirit, and for your vision. To
Severine, for imagination and dogged faith in
possibility. To the Greenhorns team and our
collaborators at Agrarian Trust, especially Lucy
Zwigard, Ian Reid, Jamie Hunyor, Kate Morgan,
Lydia Lapporte, Katie Eberle, Ian McSweeney,
and Bryan Comras—thank you for helping
to make this edition happen. Thanks to Alli
Maloney and Sue Duvall for that most essential
work of bettering our sentences. And to Tatiana
and José, thank you for being superb collaborators and for your patient pursuit of beauty
and typographical perfection.
    —BRO

DEDICATION

This Almanac is dedicated to our descendants.

THIS ALMANAC BELONGS TO

_____

TABLE OF CONTENTS

July   Wilderness · Compost   BLM Land, Wildlife Refuges

August   Arguments   Conservation Corps, the White House

September   (Built) Landscapes   Transportation Departments

October   Remediations   National Parks, State Parks

November   Sky · Vision · Seeing   Air Board, NASA

December   Invitations   Voting, Research Stations, USPS

humans woke up to the value of growing food, to the fact that a field of potatoes is the epitome of real.

Yet as vital as farmers are, they cannot single-handedly save the world. They need—as the commodity farmers surviving government trade wars thanks only to government bailouts know well—the support of people and government. More than acknowledgment, argue the writers in these pages, they need land reform, fair prices, and health care. They need respect, and they need help from urban and suburban users of the land. They need architects, wildlife refuge managers, and citizens who vote.

Because what might save the world—and by world I mean the network of wondrous green things, waters and gases, the air we breathe; the ants and worms and bees; *life on earth*—is this: all of us, ALL, heeding the advice of one of our contributors and starting to think like agroecologists, on and off the farm. From the tables where elected representatives hash out policy to the plants along our highways. From developers to mutual aid networks to exercises in environmental democracy. From backyard art to negotiations for tribal sovereignty. It is

that idea, that belief in our shared responsibility for one another and the land, that animates this *Grand Land Plan*, and that we offer as a guide for 2021 and beyond.

The path to collective survival requires reckoning and repair for our communities and the complex ecologies we love and rely on. Let's restore the fishways, our contributors say, and let's restore the oyster reefs, the forests and savannas. Let's restore land and water to the descendants of those from whom it was stolen. Let's learn all that can be learned from everyone who has survived despite—not because of—capitalism, white supremacy, and the philosophy of the #1. Instead of turning our attention away from the earth, to the high-res galaxies flattened into the black holes ever at our palms, let's look to the dirt.

Let this be the year we live into the present and prepare for a future on Earth—a future in which our planet remains inhabitable thanks to the collaboration, intelligence, and reciprocity of the humans and nonhumans who reside here.

This is our invitation to you, reader, whoever you be. ○

Poppy Litchfield   *Tide Maps*

Severine von Tscharner Fleming

INTRODUCTION

No matter what is happening in the outer circumstances, to stay open.

—Pema Chödrön

Hard to think big, while stuck inside. Stuck inside our bodies, stuck inside this rinky-dink doomsday. Each one of us feeling thwarted. Avoiding the smoke. Avoiding the corruption, the crackdown, the bailout, the backlash, and the lash-out. Such a lot of hunkering, bunkering, domestic digital addiction. A reminder how profoundly we humans ALL require access to our forested uplands, our commons, for a calming walk, to inoculate our microbiomes. To breathe in what the trees breathe out with each transpiration: love, serve, surrender.

Bodies of forest, bodies of water. Bodies at borders, in camps, on rafts, struggling to move northward from places of climate consequence. Ambition to get out, to survive, fists clenched with the willpower of migration. The faith to flee. Bodies fighting fires, bodies cutting meat, bodies hauling goods, loading ships, swiveling dollies; bodies dutifully riding up escalators from the transit system, reporting for work. Bodies making sense of what this threat to the body means, or might mean, for family, for community, for nation. Simultaneously, thousands of bodies slack, in pyjamas, not working, not playing, not flying, not up and down in elevators, not dancing or dining out. Bodies denied purposeful movement, simmering in the brain-juices of trauma and polemic. All

these bodies tell the score, rhyming with the inevitable pain of our epigenetics, our cultural identities, our appropriations, our complicity. Bodies in the streets vibrating with hurt. Bodies masked, enclosed in individual plastic packaging. All of us in little bento boxes, bound into supply chains carved up before we arrived here, cutting like wounds into the living world.

Here we are. There is no whole sane place to escape to. What is left is left to us. To do something. To discover our place in the work, in thankfulness for the gifts, in acknowledgement of the poisoning, in the humility of a human body that holds us in—and against the temptation for inaction at a time that requires us to jump into motion.

This fifth edition of the *New Farmer's Almanac* is about tuning into the needs expressed by the life force of this world we share, each in our own corner, our own basin. Each contributor to this book has in some way tuned in to "earth life"; their work, action, or thinking has committed them to the civics, forethought, bravery, stamina, and cooperation that affirms the destiny of their home place. Each is working, restoring the function and health of working ecologies, preparing home places for dire predictions as they come at us in waves. Homing in on home-making, home-keeping, home-protecting, home-nurturing. Growing greater yields, greater resilience, greater humanity, deeper safety—whether in urban gardens, fish ponds, irrigation ditches, forests, roadsides, or intertidal zones.

What is left to us is what is left to us to help heal. Acknowledging our limitations, the pain, safety considerations, and historical violence— those of us writing here, reading here, thinking and working on land, have enough traction for tremendous changes. Our grand plans may come true yet! Many hundreds of thousands more of us need to have access to land and subsistence, affinity and security. A basic income is not enough; we need a basic ecology. What would it mean if we tried to increase tenfold the actions toward resilience, the training for new farmers and restorative professions? What if we mobilized our currency to pay for it? Can the experiences and case studies and envisionings of this multigenerational movement catalyze state-funded initiatives on a far grander scale?

More of us must and are rising to serve, in office, in bold culture-making, program-making, and policy-making, and in binding the tattered kinships of each basin of relation, holding gently our togetherness, our affinity for life. The next decade will be about recruitment TO THE WORK—of makers of local change, healers of land violations, civil servants, performance artists of restoration in the commons. Our human brains have evolved for complexity, nuance, and adaptation. The work ahead is work for us.

Tidal Currents

The Roosevelt family spent summer holidays on Campobello Island, a small Canadian island at the edge of Cobscook Bay, just inside of Grand Manan and Nova Scotia on the Atlantic Coast, and just visible from my bedroom window at my farmhouse in Downeast Maine. As a nine-year-old, Franklin Delano Roosevelt (FDR) was mentored by an elder of the Passamaquoddy tribe, Tomah Joseph, with whom he explored the bay in a birch bark canoe.[1] Biographers say he gained his appreciation for water power from experiencing the twenty-two-foot tides. He fished and romped, and grew the sensibilities that would inform lifelong convictions for conservation and natural resource management. His presidency reflected that sensitivity; his institutional creations and those of his colleagues in the Soil Conservation Service, the Civilian Conservation Corps, the works Progress Administration, the Conservation Districts—including of course the thousands of dams and drainages—were attempts to merge human initiative with the natural potential of the landscape. He put millions of individuals into action to protect the underlying ecological power of the United States. Flawed, yes. But how instructive for those who would articulate an agroecology platform for the Green New Deal to study this play-out of values and outcomes.

Look out over nearby Passamaquoddy Bay when the tide is low, and you can see a herring weir on the north side, haunting as a cemetery on high ground and testimony to an abundance interrupted by "progress." The watershed of the St. Croix River drains one million acres[2] as it pours from the north into this confluence of waters. The whirlpools of these bays once surged with unbelievable life force—an extraordinary volume of herring, their wellings and swellings driving fish life, bird life, and whale life.

Homeland and fishing grounds of the Passamaquoddy for thirteen thousand years, it is estimated that historically between thirty and forty million alewives returned every year to this river, more fish than in the rest of the state combined. But in 1936, seven miles of causeway were built for FDR's tidal power project at a cost of 15 million dollars.[3] The road neatly divided the Cobscook–Passamaquoddy convergence and created a funnel to contain the electrogenerating potential of these huge bays, cutting right through the Passamaquoddy reservation. The road interrupted most but not all of the fish; this remains a place of global ecological significance and marine productivity, including some of the last 325 right whales to exist on earth. The hydrodam was never built, but still that weir stands silent; people say it doesn't really work.

Story layers on story. As we in the US enter into public works conversations for the next stimulus bills, and the next midterms, on which logic will our economic recovery hinge? Last time around, the decision was to dam the rivers. The lyrics in Woody Guthrie's commissioned songs laid clear the thinking: let the water do the work.[4] This time around, it's "clean green" energy that gets the headlines. Our addiction to electricity is ever greater with warehouse fulfillment robots and the hurtling fastness of giga-everything, with each streaming video, each bitcoin transaction, each increasingly plugged-in kid. Glowing faces of children yanked out of social contexts and plunged into flickering silos of poisonous impulse and artificial reality. A damned river drops its sediments, becomes listless, loses its rhythm. So too the children dammed up behind their screens. And so, too, a debate that swaps out the petroleum poison for another

Demonstration of bomb harpoon killing North Atlantic right whale, 1877

flavor[5] without affirming the urgent mandate to REDUCE and reshuffle our systems so that we require much much less energy, each of us.

It does all seem to hinge on how we make sense of our own safety, on what we are willing to give up and what we are willing to commit to. On which human bodies, water bodies, and cultural bodies we hold in allegiance. On whether we are able to orient our "build back better" with an ecological integrity based on systems of diversity, resilience, absorbancy, and vegetative buffering. These standoffs at Standing Rock and Pebble Mine, the latter of which risks contaminating the biggest salmon run in the world and the biggest river system of North America[6]—what could they be but decision points for our species? Does our hankering for endless disposable immediacy obviate our true duty as one species among many? Or are we willing to harmonize our human laws to match the laws of nature? Do we have the wherewithal to define a forward path that affirms ecological justice, restoration, and reparation? Are we able to kindle a guiding narrative about homeland that will help us all orient our individual vectors and volitions into a magnetic allegiance with the earth? What kind of nourishment prepares us for bravery? What does that leadership look like at the most local level? Who is going to do it—will you?

What Is Left Is Left to Us

This work will require us to tune in to the body of the earth in a different way. To heed the insights for human settlement, movement, harvest, and tending learned through thousands of years of observation by Indigenous peoples and settlers tuned to the earth's soils, drainages, and navigational potential. The Mayan milpas, Zuni waffle gardens, Spanish colonial *mercedes comunitarios* (communal

Vegetable gardens surrounding the Indian Pueblo of Zuni, New Mexico, 1873

land grants)—all these have been mapped before by careful observance. These places of potency and potential on the landscape are discoverable to us now as well; we too can map with high levels of specificity the creaturely habitats, the bird layers, the estuarine boundaries, the intersecting factors that inform the migration of animals across millennia. We have tools sensitive enough to take a more nuanced approach than "fence row to fence row." FDR made policies driven by engineers who wanted to capture the streams of potential energy, interrupting the rivers, blasting the mountains. This next phase requires us to constructively apply the tireless tally-taking of field biologists in the healing of our waterways, our forest uplands, our wetlands, our wind buffers and stream edges. It may have taken Paul Bunyan to wield his axe and fell the mighty redwood, but even a middle-school girl has enough strength to shovel a hole for a young tree.

This means cherishing and protecting our most critical places: the headwaters, the streamsides, the places where the herring swirl and can be caught, the best beaches for clams, the mountain lion corridors, the Camas Meadows. This kind of work has all been done before; Spanish arrivals to what is now New Mexico found the curve of the valley that let them channel a diversion ditch—an *acequia madre*—off the main stream of the river, a compromise with the native hydrology. The Polynesians mapped and ingeniously manipulated the upland springs to flow down into their taro ponds, where terrestrial richness became a nursery for baby fish. Those scientists who've been fed by agriculture, powered by electricity—one thing we can say is that they've made us some very potent data layers. We know the biodiversity hotspots and the Vavilov "centers of origin" where crop species arose and came into domestication; we know in minute detail the historical courses of all our water systems; we have LIDAR to describe the whole surface of the earth. We have the tools to approach this project of healing with tremendous biological integrity and an efficiency of effort as orderly as CAL FIRE, but more proactive, and less brutal.

The key ingredients, of course, are political will and cultural consensus that we humans must constructively protect our living systems, that our best farmland cannot be developed, that our headwaters cannot be clearcut, that the earth is a commons we must share and steward for the long term. We know enough to know where we need to start planting, though we cannot predict which of us will be hosting and which fleeing as the islands of habitability shift and flux. As more and more of us discover ourselves in the cause of healing, will we reconstitute the kinds of social relations that hold us accountable to the many lives bound up with our own?

Safe Harbor

Most harbors, these days, are not actually safe— not to eat the fish from, anyways. But there is safety in a group, and I'd argue that the harbor is a potent grouping to organize from. Bound up in the harbor's cultural bundle lie concentrated the human activity, settlement, sediment, trade, wharfage, anti-fouling paint, runoff fertilizer, and effluent—and also a coherent constituency, a posse with a reputation to defend. In the 18th century, our Maine trading towns and river mouths were known in Canton, Shanghai, and Cork for the magnificent boats, schooners, and clipper ships sent out into the world. Sheltered river mouths made possible the fur trade, the lumber plunder, the settlement along tributaries. With a restoration lens, we can approach each harbor as it has defined itself, addressing each particular history though the ballast of meaning and interest held by the first Indigenous settlers, the propertied early trader families, their descendants and associates, the migrants, the ethnic neighborhoods, the condos, the wharf-rats, all who are there. It is, thankfully, a defineable "we" with a group identity—the kernel of an activation.

The contamination of the harbor concerns a defined land-slice of humans. This drainage is not a metaphor but a specific place where specific people would like to be able to fish again, with confidence. Terrestrial transpiration accounts for 85 percent of rainfall on land. The paving, plowing, and deforestation of the water basin drive dessication. As our basin loses water cycling, it dries the neighboring regions as well, driving a net loss in carbon biomass—

Matt Biddulph   *No Dumping, Drains to Ocean*

less water, less plant growth, less leaf litter, less soil-building. The sponge withers.

The "act locally" crowd has long understood that watershed boundaries can orient our cleanup. They already orient the people stenciling fish skeletons and "Drains to Bay" on street gutters, labeling streams, stitching creeksides with willows, widening culverts for fish passage, and rushing out in the rain to protect salamanders as they crawl across highways.

Can we make this restoration logic catchier? Can we translate and transpose from the logic of transportation and navigation to that of transpiration, evaporation, respiration, and precipitation? Could *Sesame Street* put in some curb gardens and sing songs about the water cycle and help children follow the water upstream like a fish to find the first blockage, the first point of intervention? What will it take to un-block ourselves for this work?

Dam Site

The first dam on the Pennamaquan River was at the old Pembroke Ironworks. This was a typical 18th-century stone-built industrial compound made by Welsh, Irish, and English immigrants, quite grand. Ores brought in by sailboat were melted down by forges fanned with water-spun belt blowers and made into the iron pins and shanks and various nautical buckles and hardware for sailing ships. Shipbuilding was a major activity on the Pennamaquan River because it is protected and has a nice wide flat sandy bottom. So, the working of iron caused the damning of the river and the interruption of the yearly alewife run.

What was once a densely settled industrial district—a village along the river—is now overgrown with wild hops, wild chokecherry, eastern poplars, jewelweed, willows, muskrat, and chirping little parulas who nest in the usnea-covered apple trees along the forest's edge. This is a place that has seen nearly one hundred years of out-migration. Fewer schools of fish, fewer schools of kids.

Grey Lodge stands tall at the site, wrapped for winter awaiting her clapboards in springtime (yes, we need help with this project)—a noble-boned boarding house with a brand new roof. We Greenhorns got it at a tax auction and have set it up as a fermentory and hospitality center, fermenting feral apples into apple cider vinegar and welcoming conservation-minded tourists to our little town on the coast of Maine. Prior to us, the building was used by Golden Hope Mines, a mining speculation company that drilled into mountains and doused the cores with acid, trying to find gold. We inherited a basement full with sodden mining ores. Funny the psycho-geography of this particular rise, where the dam site causes a whole watershed to pool up underground, flooding our basement. As the well man said, "You could start a bottled water company here!" We've put our mycological lab into a room with mining procedures scrawled in marker on the walls— it's nicely sterilized now. Next is the trash room, then the ceramics studio.

Beside the dam is the new fishway, a simple

→ *Men Drilling for the Passamaquoddy Tidal Power Project, 1936* ramp that lets the fish jump up a series of ladders to get upstream to spawn in the warm lakes. We made a film about it called EARTH-LIFE: FISH, so you can watch and read online about how Downeast Salmon Federation engineered this ramp to help the alewives back over the dam in larger numbers, perhaps up from seven hundred thousand to three or four million returns per season. Between Grey Lodge and the dam is the Crossroads Motel, long rectangular trailers of moldy vinyl rotting into the hillside—a place of tremendous potential as a commons if we can get the price reduced and some help with redevelopment.[7]

Funny, the quirks of retrograde opportunism in our down-and-out neck of the woods; no sooner had we pumped out the basement than the miners came back and wanted to claim the ores that they'd abandoned for years. Now that the price of silver is up, they'd like to reprocess those ores. The guys around town say this would represent more than a million dollars in exploration costs. Maybe we'll let them haul it all out to make more room to store vinegar. Or maybe we'll use the ores for drainage in the facility we build to process organic wild blueberries. Mother mountain, will she be mined to make more transistors and gizmos? Are we to have another smelting event in this watershed?

How can we repurpose the buildings and materials and institutions we have been left with? Can we turn the cemetery into a fruit commons? the historic fort along the river into a grazing commons? the Odd Fellows hall into a community café, a place for Al-Anon meetings?

Can we reclaim the main street abandoned for bigger towns and malls? remake it as a place for daycare, beauty parlors, a local food depot? This adaptive reanimation frames our approach to the work ahead. We admire those grand old textile mills in Biddeford turned into bakeries and commercial kitchens, and we're glad that the wharf-buildings were saved and turned into high density condos. But it's pretty clear that whale watching and summer tourism will not keep our rural schools from consolidating and shutting down. Nor can we reopen the twenty-seven canneries that have closed along our bay. We have to figure out things to do here that can endure, with settlement patterns and employment that suits the circumstances we find ourselves in.

Instead of canning fish, can we can fruit? Instead of smelting iron, can we smoke and ferment fish? Instead of an economy based on exploitation and export of raw materials (wood pulp, biomass) couldn't we rebuild a conservation-minded, high-value, natural-resource economy based on what grows here in wild abundance? The blueberries, the sponta-neous feral apples, fruit adapted and adapting, expressing the free will of self-selection, the wily ones surviving. The half-breeds that handle the late frost, the early frost, the late spring, the early spring—year after year. In historic preservation circles as in horticultural ones, there's a tendency to miniaturize and fetishize the "heritage" apples chosen for baking, pie, and sauces. Valuable history, yes, but a house museum will not suffice for the future we're facing. It's time to cherish the half-breed apples, the cross-cultural culinary potential of this landscape in a great fullness of the enterprises and microproductions that can work here. The sweet fern, sweetgrass, sweet bay, balsam fir, goldenrod, rugosa rose, St. John's wort, wild mushrooms, and wild honeys. Just as remote mountain villages in Romania, Switzerland, Albania, and Spain support a supply chain of aggregators and processors for such products, couldn't we in Washington County dry and distill our wild glory and ship it to market?

This is an argument that goes far beyond farms and food, but let's start there, with a delicious umami broth made from our Maine nori and kombu, Maine-fermented fish sauce, Maine-fermented miso, Maine shiitake, Maine wild chive blossoms, and Maine carrots and leeks and caraway seed.[8] What a powerful kaleidoscope we can reconstitute from apples from the forest's edge, analog climates, shisandra and szechuan peppers, quinces and chestnuts, rhodiola and greenhouse ginger, wild natives and wildish forest-scapes of our own invention. Hardy kiwis, let us see what these twirling trellis-lovers will yield!

Softening the Landing

As above the waterline, so below. Before we leave the waterfront, I want to tell the story of Sunken Seaweed. They're a group working to clean up contaminated waterways in San Diego. On shaky old wharves, they grow a seaweed called ulva that metabolizes phosphorus and

Wild Maine blueberries

nitrogen, binds heavy metals, and takes up carbon. This seaweed can then be made into biochar or compost that can be used to remediate urban soils, to fertilize young urban trees surrounded by woodchips and swales planted with native bushes that trap dust, sediment, rainwater runoff, and contaminants. It can help remediate waterways degraded by the port, the military, the suburbs, and the cities, healing the human footprint on a delicate ecology. Along with the structural microcosm of the biochar, the hormone-laden seaweed creates the conditions for more life. The soil becomes more porous, more spongy, more filtering, more holding of moisture. This is the land-softening that is needed upland of every harbor, every waterway, and particularly those so degraded.

Many harbors have port authorities keen to ameliorate their working waterfront. These are places which have quite a lot of jurisdictional agency and potential for federal funding and innovative reuse. The Port of San Diego happens to be an entity with lots of land and cash to do some experimenting. Such organizations, like hulking hulls of tankers, can support many young barnacles, many young remediators. These small initiatives can instigate a succession of enterprises, especially if they are conspiratorial and not overly concerned with the hype of venture capital. If they build kinship and share insights, connect with municipal planting departments and climate change researchers. If they reconnect in the way that each place must.

Work Is Essential

If we are serious about resilience and adaptation for the whole of our human society, then we must be serious about the work that is involved. Will this new stimulus program move the work into places where work is needed,

where new businesses and organizations are needed, or will it all stick to the trillion-dollar ribs of the largest dinosaurs in the US economy? If we're serious about restoration, we need to make this work available and accessible and remunerative. We need to create trainings, stipends, grants, loans, and professional pathways. We need a diverse range of economic forms within which to work (e.g., landscaping companies, restoration contracts, sideline work, seasonal work). We need this language of possibility and this physical work made available to those who would enter the Peace Corps, the Conservation Corps, and Americorps; to those who would enter professional sports; to those who would enter the armed services; to those who are already working in the fields. We need massive structural reform in creating REAL OPPORTUNITY for training and pay in the restorative arts. Imagine blending the moral indignation of the Edible Schoolyard kids and the climate strikers—the Gretas and Extinction Rebellion and the Sunrise Movement—with the out-of-work granddaughters of miners and lumberjacks, the sons of liberty, and the kids with vocational degrees. Imagine them all getting oriented to careers in restoring ecological infrastructure. It happened during the Great Depression; millions of working arms were mobilized by the Civilian Conservation Corps, fed three square meals, and provided lodging and vocational training for massive public works projects. Imagine the side-by-side working, not of a prison crew, but of a mosaic social unit reflective of who we are as a nation and a species.

Many of the places most in need of restoration in the US are the places where disinvestment, contraction of industry, consolidation of natural resource economies, and population decline have coincided with fearfulness and retrograde identity politics. What if there were well-paid work planting street trees in these towns and cities, tending hedges around those mill sites? What if the strategic beautification of our degraded places were so successful that these became highly desirable places to live once again? What if doing that work alongside people who are different from ourselves spurs the regrowth of compassion and neighborliness?

We need an applied, sustained epoch of epic loving handwork—a multidimensional, multi-generational project of earth repair. Work that is available at all scales and to many kinds of people, many of whom will need to be trained. It must happen in a manner that affirms the economic and cultural diversity of our urban and rural places. There have been calls for a Green New Deal. I'm hoping that federal funding and stimulus packages that follow on the heels of the US CARES act comprise after-school programs, job training programs, educational stipends, and funding to states for grants and conservation innovation.

Moving Out to the Country

It used to be the punk rockers in Detroit who put bumper stickers on their bikes reading "Fuck cool cities"—now it seems even the cool cities are fucked. People are moving. Some are moving to small towns because they can work from anywhere. Some are moving back to small towns to attend to aging family members or take over a family business, or to shelter in place with lower overhead. Yes, there is gentrification going on. We in the countryside can all see it and feel it, the slice-and-dice of

❋❋❋

What is leadership?
* Take responsibility for what you know needs to happen.
* Put yourself in a position to create impact.
* Think from where you sit and what you know: What would it take to develop the needed skills and leverage points? What would need to change? Who else would need to be involved and mobilized?
* Articulate a plan.
* Line up the people and convene the necessary conversations.
* Make it accessible at the lowest possible level of governance.
* Act like you've already won.

* Buckle down and commit to the social work it takes to make the ecological work work.
* Figure out how to learn as quickly as possible to iterate.

Multiple Choice!

Public trust	❋	Public confidence
Public works	❋	Public office
Pubic health	❋	Public land
Public service	❋	Public water
Public interest	❋	Public rights of way

❋❋❋

sanctuary making—of whatever flavor of vacation home—that hacks into farmable fields. Harder to figure out is how to create the conditions where people begin moving to, and moving back to, small towns specifically to undertake the kind of work that makes the nation stronger, healthier, and more absorbent of rainfall.

What's the difference between someone who's coming to engage versus some who's coming to hide away? Since the fall of the age of sail, many Maine islands and coastal historic towns have become summer-home communities, without the hospitals, firefighters, and civic participation to actually cope with the summer-flush. Consider the small-town volunteer fire department's fourth of July parade and picnic, with hot dogs and potato salad, iced Bundt cake and styrofoam coffee cups, flags flapping over the hot afternoon pavement. The ladies auxiliaries of the Grange, Odd Fellows, Veterans of Foreign Wars, Elks, and Lions have been hosting such iconic town happenings for decades, many of them into their eighties and nineties. Fraternal orders, town governments, volunteer firefighting, rural schools, rural churches, rural foodbanks— these institutions share the crisis of contraction and out-migration.

Here's the thing: Towns take work. More than many people feel they have time for, and the current crew have been holding things together for longer than is realistic already.

In many cases, or so goes the scuttlebutt in farm circles, the re-ruralizing underway isn't playing out in sync with actual civic needs, civic services, or civic engagement.

Towns require townspeople, people who take on the PTA, the council positions, the bakesales and bean-supper coordination, the recycling depot at the transfer station, the conservation district board, the planning board, the historical society, the grantwriting, the watchdogging, the citizen science, the neighbors helping neighbors. City-raised people don't always have familiarity with these practices, this tradition of volunteerism. Weed growers, wine growers, horse people—they get a bad reputation as takers, and sometimes for good reason. This is a fixable issue, a potent transformational tactic for the coming decade— to reenergize these local positions and revalue the humility of working together.

Building back better doesn't just mean big-money school bonds, or big-truck contractors fixing the bridges; it also means reenergizing our small-town institutions, our county supervisors and town councils and volunteer fire departments. It means fostering networks, fostering the caring that it takes to hold together. This is a discourse I look forward to in 2021.

It's been perplexing to watch as the "have alreadys" gobble up real estate from the "stillhaves" while so many feel relegated to "never will have" status. Efforts to create social finance to support land access and equity for new farmers, farmers of color, and young tradespeople, like those of Agrarian Trust, struggle to keep up as land prices climb and climb. As my "cli-fi" reading friends tell me, a few more pandemics could produce a far less populous world where chutzpah, skill, and teamwork would be more powerful than capital.

Those countless forlorn main streets in our vast land invite re-invention, if the invitations go out. The hazard of expecting the work to be done remotely—by a robot, by someone else— is that the work often goes un-done, the newly burned forestland naked in the first strong rain, running off into the creek. As I sit here full of hyperbole and conjecture, the ecology continues to unravel.

Roadsides and Radical Gifts

This road we're on seems like a crash path, but there are still points of intervention. Barbara Deutsch, my eighty-year-old fairy godmother, and I were en route to our first butterfly walk together, driving in the Oakland hills, she like a safari driver zooming around sharp corners, swerving as she called out the ceanothus and pipevine hedges. She made rounded gestures and cooing sounds in reference to the sun-warmed south-faces where the lepidoptera love to alight. Amidst the golden glow, we came across a stretch of road littered with trash. I made some silly comment about people who throw trash out windows. She hit the brakes, slowed nearly to a halt, looked me straight in the eye, and said, "The trash isn't the issue, Dear. The road is the issue."

This road we're on is not a metaphor; it is a road. But the road is not only a road; it is also a water catchment system, a micro-dam, a solar collector, and a pollution vector. The road exists in multiple dimensions, as do we all. The road is a design challenge, a site for the public art of erosion control, for native plantings, mulch-bag garlanding, and water-catchment demonstration. The road is not just a violation; it is an opportunity for remediation and redemption! This road was built by the Works Progress Administration during a great depression.

Each curve provides a wrinkle of drainage that can sustain a very happy native tree or shrub. Barbara's other fairy goddaughter, Amber Hasselbring, managed to negotiate a necklace of habitat across San Francisco for the green hairstreak butterfly. She connected patches of habitat in parks, creating a tender, dappled flutter-path across paved neighborhoods. The work is not glamorous, but with her twinkling smile she offered to rip out the pavement and plant habitat. These interventions in the road give us all a place to slow down, a place for the collected debris and dust to settle, a spongy, mulchy spot to slow the water and sink it. A place where animals can cross. A place to advertise to one another that we care about beauty. That we care about each other. That love is possible.[9]

George A. Grant, NPS *Civilian Conservation Corps enrollees carrying on erosion control operations along Vicksburg National Military Park's Graveyard Road in 1934*

Shade Trees and Group Work

What actions and cultural approaches affirm human survival? What kinds of humans do the trees need us to become? Common to the documented functioning of a commons are seasonal, ritual, public rites of engagement. Be it the springtime walk (and cleaning) along the contoured irrigation ditch (*la limpia de la acequia* in the Southwest US) or descending the mountain meadows in a formal costumed procession with cows wearing crowns of flowers (the Swiss alpine commons) or visiting the water temples at the top of the watershed (Balinese rice gardens), these cultural happenings integrate the janitorial duties associated with managing large-scale ecologies, agroecologies that sustain whole systems. They make clear our affiliation to the land, our belonging to a place, our connecting to one another, and our responsibilities to the whole. These boundaries and the bonding that goes along with them create consonance, aligning each individually charged needle with the magnetism of the entire watershed.

New Zealand is a land with no native mammals, a sheep colony that exports spring lamb and milk powder through Chinese-owned ports—but it is also a place where the pregnant prime minister announced a campaign to plant a billion trees the first week she was in office.

A producerist, maximalizing behavior has come to define much of New Zealand's agronomy. I've been told the rapid removal of subsidies in the 1980s led to this massive intensification and export optimization. Fence posts are dipped in arsenate, ditches are sprayed with herbicide, airplanes drop baited poison, and far too many cows stand beside polluted rivers in fields far too irrigated, fertilized, and compacted. It's enough to give you toxic pinpricks, this puncturing and neutering of the microhydrology, this interruption and violent reorientation of the sentience of the landscape. Bulldozed new road expansions, on-ramps ripped from delicate dunes—land violations to the left and the right. Then the land slumps and collapses, and scarred hillsides are bandaged up with a sad little repertoire of European fix-it trees. Poplars twinkle, but really it should be bush. It could be bush. Thank god for manuka honey, so that some still is.

As usual, that is not the whole story. The mainstream is not the whole of the watershed, and thanks to the Edmund Hillary Fellowship, I was able to meet and experience a powerful subculture of activated, regenerative farmers and thinkers who are transitioning their

Kai 'Oswald' Seidler   *Rice field in Bali*

pastures and operations. Intensifying grazing management, reseeding paddocks with dozens of clovers and diverse forages, restructuring their farming such that the land can fix its own nitrogen and the farmers can kick the habit of importing rock phosphate from across the world. Is such a transition as daunting as clearing the forest to begin with? Surely not—if those ole' colonial bushwackers and lumberjacks in the 18th and 19th centuries could build these dams and knock down these forests, we can muster the willpower to build back better!

As I sped along to meet the wonderful protagonists around the tidy countryside, I kept noticing the long stretches of roadway lovingly planted with charismatic native plants, somewhat in the style of a 1970s hotel landscape, but in a good way. Dramatic six-foot hedges of New Zealand flax, flanks of reflected light beaming off the vertical panels. And below, twinklings of the heather-like, the fern-like, the sword-like, the upright reeds, the gunnera stalks, pitospernums, the furry paws, the privet-like, the box-like, the unfurling waving wands, the pom

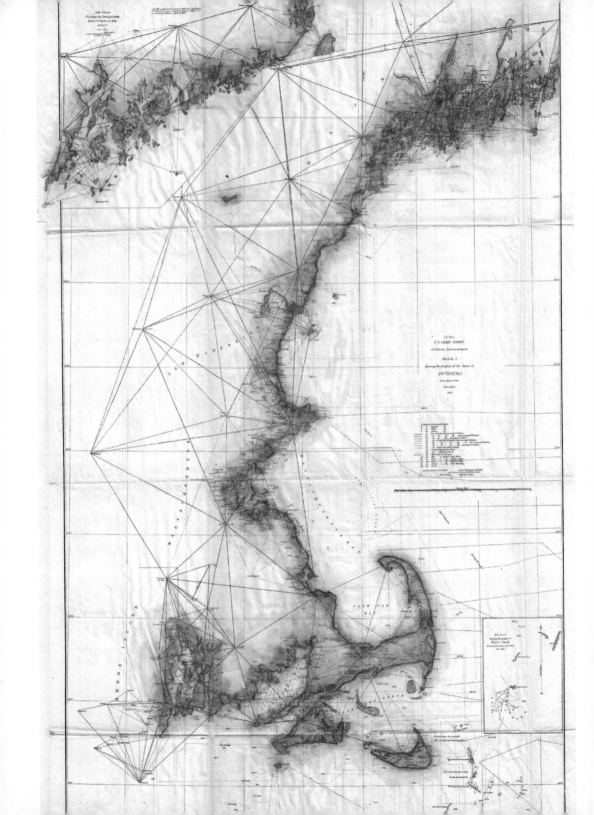

poms, tendrils, cups, and lilies. On and on! Such a marvelous Pacific island flora, blacks and purples, temperate and sub-tropical. In quite a number of provincial regions, planting contracts are held by Māori organizations, supporting solid jobs. Even if this highway treatment amounts to a Potemkin treatment for tourists, I'll take it. Let's start there—someone get Greta a shovel, and let's bring those fierce transgressive teenagers out to reclaim the commons, commondeer the center median. From there, they'll penetrate other boundaries, other constructs, other wounds that need tending.

Because I agree with them—it's not fair. We "adults" should be ashamed that the young have to contend with and content themselves to coping with such a degraded planet.

As the trillions trickle out, it seems imminent that we'll realize that only within a reviving ecology will our aspirations for recovery work out. And that swinging a pick to plant a shrub, stitching back the walls of a creekside, and restoring the cultural infrastructure are what make rural living convivial. What politician could oppose a public works project housed in a community college or trade school, offering paid training in restoration and afforestation? What congressman would oppose native plant nurseries with state contracts and professional training sites at high schools? What mayor could ignore an urban forest strategy that actively increases air quality, cutting up pavement to make way for drainage, infiltration, windbreaks, shade, and cooling? Why not riparian restoration to house the birds and pollinators along all the flyways? bioremediation of lead- and metal-contaminated urban lands? Buffers, fruit trees, gardens, and drought-resistant native flowers where now are lawns! Church, school, and community gardens, dust-trapping hedges that bear fruit! Grants for towns to invest in historic preservation, hire local contractors, restore civic buildings, churches, and community centers! Grants to remediate parking lots! make biochar! tend forests! build and maintain hiking trails!

The list is long and grows longer with all the voices contained in this Almanac. Each expression of hope, each analysis of the work that lies ahead, each poem and essay is an invitation to you, our public, to collaborate with the public thinking and policy that lies ahead. Merrily we must repair and revitalize, adapt and renew, restore, refurbish, reconstitute, and shore up for what's still coming; to see how well we can outlast the storms. ○

Notes

1. That exact canoe now sits in the Smithsonian Museum, lovingly restored by Steve Cayard and David Moses Bridges, an almost lost art which they have since revived. Come learn to build one during Greenhorns summer camp programming in 2022, COVID-pending.
2. California's first one-million-acre wildfire burned this year, and was dubbed a "giga fire," so in Downeast parlance you could call this a "giga river."
3. *Passamaquoddy Tidal Power Project*, curated by Christina DeBenedictis for the US National Archives, n.d.
4. Woodie Guthrie wrote "Grand Coulee Dam" and twenty-five other *Columbia River Ballads* when commissioned by the Bonneville Power Administration in 1941. The song promoted hydroelectric power with lines like, "Roll along, Columbia, you can ramble to the sea, But river, while you're rambling, you can do some work for me."
5. President Obama's green energy "beyond coal" mandate turned out to be a boom time for natural gas. See *Planet of the Humans* by Michael Moore for a takedown of the green jobs/green energy boondoggle.
6. Richard Read, "This Alaska mine could generate $1 billion a year. Is it worth the risk to salmon?", *Los Angeles Times*, October 23, 2019.
7. Grantwriters and social investors sought to partner on this project.
8. This "Mermaiden broth" is available at smithereenfarm.com.
9. Inspired by this work we Greenhorns have created a FREE guidebook called *Habitat Everywhere*.

← **A. D. BACHE, USCS** *Sketch A Shewing the progress of the Survey in Section No. 1*
The US Coast Survey's 1859 triangulation chart of the New England coastline. An inset map in the upper left focuses on Penobscot Bay.

Jon Levitt   *Parula*

✳✳✳

Postscript:
Here's a song recommended to me by Sharifa
Rhodes-Pitts, whose beautiful voice I hope you
will hear when you tune into the EARTHLIFE
podcast. She has written a wonderful book about
Harlem, *Harlem is Nowhere: A Journey to the
Mecca of Black America.*

Bernice Johnson Reagon's performance of this
traditional gospel song is available on *Smithsonian
Folkways Recordings.*

> Come and go with me to that land. To that land where I'm bound.
> Nothing but peace in that land.
> Nothing but peace in that land. Where I'm bound.
> No more hatred in that land.
> Come and go with me to that land. Where I'm bound.
>
> We'll all be together in that land.

✳✳✳

Celestial Calendar

PHASES OF THE MOON

	Sunday	Monday	Tuesday	Wednesday	Thursday	Friday	Saturday	Sunday	Monday	Tuesday	Wednesday	Thursday	Friday	Saturday	Sunday	Monday	Tuesday	Wednesday
JAN						1	2	3	4	5	◐6	7	8	9	10	11	12	●13
FEB		1	2	3	◐4	5	6	7	8	9	10	●11	12	13	14	15	16	17
MAR		1	2	3	4	◐5	6	7	8	9	10	11	12	●13	14	15	16	17
APR					1	2	3	◐4	5	6	7	8	9	10	●11	12	13	14
MAY							1	2	◐3	4	5	6	7	8	9	10	●11	12
JUN			1	◐2	3	4	5	6	7	8	9	●10	11	12	13	14	15	16
JUL					◐1	2	3	4	5	6	7	8	●9	10	11	12	13	14
AUG	1	2	3	4	5	6	7	●8	9	10	11	12	13	14	◐15	16	17	18
SEP				1	2	3	4	5	●6	7	8	9	10	11	12	◐13	14	15
OCT						1	2	3	4	5	●6	7	8	9	10	11	◐12	13
NOV		1	2	3	●4	5	6	7	8	9	10	◐11	12	13	14	15	16	17
DEC				1	2	3	●4	5	6	7	8	9	◐10	11	12	13	14	15

Thursday	Friday	Saturday	Sunday	Monday	Tuesday	Wednesday	Thursday	Friday	Saturday	Sunday	Monday	Tuesday	Wednesday	Thursday	Friday	Saturday	Sunday	Monday
14	15	16	17	18	19	(moon)	21	22	23	24	25	26	27	(moon)	29	30	31	
18	(moon)	20	21	22	23	24	25	26	(moon)	28								
18	19	20	(moon)	22	23	24	25	26	27	(moon)	29	30	31					
15	16	17	18	19	(moon)	21	22	23	24	25	(moon)	27	28	29	30			
13	14	15	16	17	18	(moon)	20	21	22	23	24	25	(moon)	27	28	29	30	31
(moon)	18	19	20	21	22	23	(moon)	25	26	27	28	29	30					
15	16	(moon)	18	19	20	21	22	(moon)	24	25	26	27	28	29	30	(moon)		
19	20	21	(moon)	23	24	25	26	27	28	29	(moon)	31						
16	17	18	19	(moon)	21	22	23	24	25	26	27	(moon)	29	30				
14	15	16	17	18	19	(moon)	21	22	23	24	25	26	27	(moon)	29	30	31	
18	(moon)	20	21	22	23	24	25	26	(moon)	28	29	30						
16	17	(moon)	19	20	21	22	23	24	25	(moon)	27	28	29	30	31			

PRINCIPAL PHENOMENA 2021

Phenomenon	Date	Time
Perihelion	January 2	08:50 EST
Equinox	March 20	5:37 EST
Solstice	June 20	23:32 EST
Aphelion	July 5	18:27 EST
Equinox	September 22	15:21 EST
Solstice	December 21	10:59 EST

January 2, 3 Quadrantids Meteor Shower

April 21, 22 Lyrids Meteor Shower

May 4, 5 Eta Aquarids Meteor Shower

May 26 Total Lunar Eclipse 4:47–9:49 EST
A total lunar eclipse occurs when the Moon passes completely through Earth's dark shadow, or umbra. During this type of eclipse, the Moon will gradually get darker and then take on a rusty or blood red color. The eclipse will be visible throughout the Pacific Ocean and parts of eastern Asia, Japan, Australia, and western North America.

June 10 Annular Solar Eclipse 4:12–9:11 EST
An annular solar eclipse occurs when the Moon is too far away from Earth to completely cover the Sun. This results in a ring of light around the darkened Moon. The Sun's corona is not visible during an annular eclipse. The path of this eclipse will be confined to extreme eastern Russia, the Arctic Ocean, western Greenland, and Canada. A partial eclipse will be visible in the northeastern United States, Europe, and most of Russia.

June 24 Full Moon, Supermoon 14:40 EST
This full moon was known by the Algonquian tribes in the northeastern US as the Strawberry Moon because it signaled the time of year to gather ripening fruit. This moon has also been known as the Rose Moon and the Honey Moon. This is also the last of three supermoons for 2021. A supermoon occurs when a new or full moon nearly coincides with perigee, the closest that the Moon comes to Earth in its orbit, leading the Moon to appear slightly larger and brighter than usual.

July 28, 29 Delta Aquarids Meteor Shower

August 11, 12 Perseids Meteor Shower

August 22 Full Moon, Blue Moon 8:02 EST
This full moon was known by the Algonquian tribes in the northeastern US as the Sturgeon Moon because the large sturgeon of the Great Lakes and other major lakes were more easily caught at this time of year. This moon has also been known as the Green Corn Moon and the Grain Moon. Since this is the third of four full moons in this season, it is known as a blue

moon. Most tropical years—years measured from one winter solstice to the next, as in the old *Maine Farmer's Almanac*— contain twelve full moons. But periodically, a tropical year contains thirteen full moons, so one season has four full moons instead of three. Blue moons occur on average once every 2.7 years.

October 7 Draconids Meteor Shower

October 21, 22 Orionids Meteor Shower

November 4, 5 Taurids Meteor Shower

November 17, 18 Leonids Meteor Shower

November 19 Partial Lunar Eclipse 1:02–7:03 EST

A partial lunar eclipse occurs when the Moon passes through Earth's partial shadow, or penumbra, and only a portion of it passes through the darkest shadow, or umbra. As it moves through Earth's shadow, part of the Moon darkens. The eclipse will be visible throughout most of eastern Russia, Japan, the Pacific Ocean, North America, Mexico, Central America, and parts of western South America.

December 4 Total Solar Eclipse   07:33 EST

A total solar eclipse occurs when the moon completely blocks the Sun, revealing the Sun's beautiful outer atmosphere, known as the corona. The path of totality for this eclipse will be limited to Antarctica and the southern Atlantic Ocean. A partial eclipse will be visible throughout much of South Africa.

December 13, 14 Geminids Meteor Shower

Data drawn from NASA Map and Eclipse Information, the US Naval Observatory, and the 1937 *Maine Farmer's Almanac*.

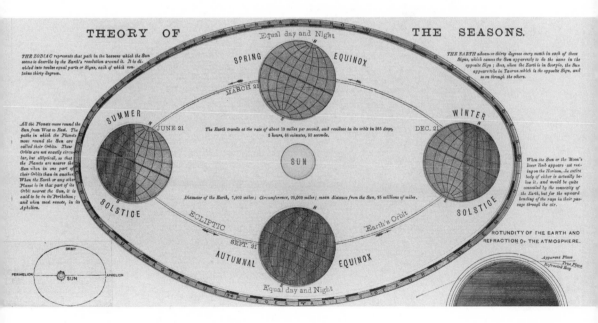

From the *New Ideal Atlas*, 1909

↑ **Donald E. Davis, Don E. Wilhelms** *Maps of the surface of the Moon*
→ **Andrew Stuart** *Great Orme Kashmiri goats on the streets of Llandudno, Wales*

Resistance · Recovery

FOIA, BIA, National Heritage & Monuments

Jenni Monet

PENOBSCOT MILLION

Colonization's Continuum of Indigenous Lands in Maine

In the fall of 1980, teenagers Kirk Francis and Mark Chavaree watched their Penobscot Nation elders make national headlines from Maine. Some media accounts labeled the tribe's historic land claims settlement a Native rights victory; others called it the surrender of tribal sovereignty. Four decades later and upon Maine's bicentennial, now-Chief Francis and Chavaree are tribal leaders—and they are still sparring with the state, a continuation of a two-hundred-year-long tussle over who owns the land: the Penobscot people who have spent lifetimes along the Penobscot River, or the colonizers who purchased the territory illegally?

Today's land dispute centers on portions of the Penobscot River that flow through the tribe's reservation boundaries. Francis and Chavaree are among an estimated 2,400 tribal citizens who contend that rights to the riverbed belong to them, an argument advancing at a particularly advantageous time in the United States. Since the Indigenous uprising at Standing Rock, the pendulum of justice increasingly favors federal Indian law and policy. Historic treaties signed between tribes and the US are being recognized for the legally binding pacts that they are. More importantly, they're being upheld. The recent landmark US Supreme Court case, *McGirt v. Oklahoma,* is a similar land battle and the most immediate victory of its kind. The fight for the Penobscot River will likely be decided by a federal appellate court before the year is out.

For Francis and Chavaree, their stake in the river war is generational. Fifty-one-year-old Francis recently secured his fifth term as Chief of the Penobscot Nation, making him the tribe's longest-serving leader since elections were first held in 1850. The son of a longtime tribal councilman, Francis followed in his father's footsteps, first serving on the council before becoming chief in 2006. In June 2020, Chavaree, who is fifty-eight, marked his third decade as the tribe's in-house staff attorney.

Born and raised on Indian Island in the heart of the Penobscot Nation, Chavaree is the grandson of fluent Penobscot speakers, the language of the *Pa'nawampske'wiak,* or "people of where the river broadens out."

Their forebears, before signing a 1796 treaty, had explicitly claimed the Penobscot River as theirs. Ancestral Penobscot territory is coterminous with the entire Penobscot watershed, curving around Indian Island like a beaded necklace, spanning some seventy miles northward toward the Canadian border and flowing thirty miles south to a smattering of islands in Penobscot Bay. Comprising four branches, the total system stretches an estimated 265 miles, making the Penobscot the longest navigable waterway in Maine.

In the Penobscot way, their namesake river flowed from a monster frog defeated by Gluskabe, the Wabanaki spiritual chief.[1] The frog forbade the Natives the use of water until he was killed by a yellow birch tree cut down with Gluskabe's ax. As the creation story goes, all the branches of the yellow birch tree transformed into water that became the four branches that today empty into the Penobscot's main stem.

Ancestral Chief Attean Elmut proclaimed in 1807 that "the God of Nature" had gifted the tribe with abundant resources along the

Penobscot River.[2] It was their lifeline. Eels were speared as they burrowed in the winter mud, and by summer, salmon were netted as they swam upstream. Guided by harvest moons, families set up seasonal camps along the shore, tapping maple trees in the spring and snaring muskrat in the fall. But in the years leading up to Maine statehood in 1820, eager white settlers built the first of many dams on the Penobscot River, gravely disrupting the tribe's annual fish run. Elmut felt particularly responsible for Penobscot families and took this invasion personally.

"Put an immediate stop to the English encroachment on our rights," he pleaded with treaty negotiators in Boston.[3] He explained that their 1796 agreement with the Commonwealth of Massachusetts—which then encompassed the District of Maine—had secured vital sustenance fishing rights to the Penobscot Nation. These rights were being violated, he insisted. But his repeated complaints were ignored. Concerned that his desperate attempts to restore the river and the fish run appeared futile, the chief dramatically took his own life.[4]

Elmut's suicide was emblematic of the grave outcomes treaty-making would ultimately deal to tribes. Chroniclers of the post-Revolutionary War era described the negotiations as "dishonest" and "vague," to the detriment of the Wabanaki nations.[5] When the Penobscot ceded an "indefinite extent" of land to the Commonwealth in the Treaty of 1818, the historical record questioned whether the tribe's terms had been properly conveyed and whether the Penobscot fully understood the agreements that they were signing. Eventually, these pacts came to be regarded by Natives as "bad paper" because every one of the roughly 375 treaties that tribes signed with the US had been abrogated.[6]

Most Americans know very little about the treaties that functioned as the mandated gateway to colonization. One of the first laws passed by Congress, the Indian Non-Intercourse Act of 1790, made it criminal for prospectors to acquire Indigenous lands without a formal treaty made with tribal nations and ratified by the federal government.[7] But even with such protection, which also promised entitlements to Native Americans, they fell vulnerable to an arrogant settler society that operated above the law and undermined tribal sovereignty outright. The Penobscot and their joint Wabanaki Confederacy were among the first to be burdened by such outcomes.

Upon separation from the Commonwealth in 1820, Maine assumed "paternal control" over the Confederacy and authorized the sale and lease of tribal property, often without the tribe's consent. Over the centuries, this led to more dams along the river, contamination from toxic paper mills, and the near-extinction of the Penobscot's ceremonial salmon run. It wasn't until another tribe in Maine, the Passamaquoddy, argued that they had been wrongly dispossessed of their Indigenous homelands that the imbalance of colonization became exposed.

In the 1950s, a tribal member retrieved from an old trunk a 1794 treaty that revealed the cession of Passamaquoddy land to the Commonwealth was never ratified as required by the US government.[8] The Penobscots realized their treaties were also null and void, and banded together, bringing forth a stunning land-claim dispute. It addressed their rightful title to twelve million acres, or roughly two-thirds of Maine. The tribes received $81.5 million to drop the entitlements—the very deal that Francis and Chavaree observed forty years ago as teenagers.

"Originally we thought that the settlement was a good thing," said Chavaree, who said his family, like most other Penobscots, were financially poor. "But the way things have unfolded, the state has taken a strict dominion over us." Today, Maine has sweeping authority over tribal affairs, an unusual outcome of the 1980 Maine Indian Claims Settlement Act. Unless granted by Congress, tribal nations are generally free from state control. Of the 574 federally recognized tribal nations, very few are subject to state laws within their territories, as those situated in Maine. The majority of tribes govern directly with the US.

As Francis and Chavaree observed Maine's bicentennial, they were reminded of the state's relentless control over their tribal destiny. The Penobscot Nation see Maine's lasting authority as a direct assault on their right to self-govern, an encroachment on their cultural lifeways and especially on their namesake river, the source of their creation story. "It's the identity of the Penobscot Nation," Francis told me from the tribe's headquarters on Indian Island. "The Penobscot River runs through a territory that the Penobscot have understood to be ours since forever."

The Penobscot Nation argued their case against Maine before a federal appeals court in September 2020. Knowing that Governor Janet Mills strongly opposes the tribe's claims that it has rights to the Penobscot River, a frustrated Francis addressed a state task force in late July by Zoom. "The tribes in Maine have been stuck decades behind other tribes in the country," the chief said, set against a digital backdrop of the Penobscot River. "Can we see a day where we're not fighting constantly on whether tribal members can fish in their own Indigenous waterways?" His question hung as heavy as the acrimony that, for two centuries, had always

Meagan Racey, USFWS *Veazie Dam Removal*
Joseph Dana of the Penobscot Nation paddles a traditional birch-bark canoe past the breaching of Veazie Dam, which had blocked the natural passage of fish on the Penobscot River for two centuries.

seemed to linger over tribal-state relations. But just as Chief Elmut had assumed immense responsibility for the Penobscot Nation and its ties to the Penobscot River, so too does Francis. The lawsuit looming in the fall of Maine's bicentennial was the latest chapter laying bare the storied colonization of America.

～

Land was an ideological priority for white settlers and resulted in the seizure of 1.5 billion acres of tribal territory from 1776 until present day. In 1796, pressured into paying off enormous Revolutionary War debt, Massachusetts sought revenue from the sale of valuable "public lands"—Penobscot lands—in which General Henry Knox, the first US Secretary of War, would close his first major deal. Recently retired, Knox had been courting British financier Alexander Baring to purchase a vast tract from the "Penobscot Million," the general's epithet for the one million acres of lush lands he obtained by lottery in the regional timberlands of the Penobscot River.[9] In a dispatch to his Amsterdam investors, Baring elaborated: "The Penobscot is one of the finest rivers in

America and its banks will become the center of population in Maine."[10] In February 1796, he bought 1.2 million acres along the Penobscot for £107,000 (approximately $3.5 million, today). The roughly 200,000 acres considered the most desirous were the tracts along the Penobscot River claimed by the Penobscot Nation.[11] By June, the Commonwealth of Massachusetts, encompassing the District of Maine, finalized the sale by treating with the Penobscots.

At first, the Penobscots had been reluctant to negotiate with the Commonwealth. For twelve years, they refused to cede land to the Massachusetts government after promises failed to materialize with the Provincial Congress in Boston. When the Penobscot became the first tribe to support the American Revolution, George Washington promised that the new Republic would protect their Indigenous homelands. But it did no such thing. Following the Revolutionary War, political authorities did little to stop white poachers from squatting on Penobscot territory. Rather, the encroachment emboldened the Commonwealth to presume its own authority to "extinguish Indian title."

It took four attempts to persuade the Penobscots to enter into the Treaty of 1796. In this persistence, one thing was evident: the Commonwealth and speculators like General Knox understood quite clearly the legal concept of "extinguishment": the explicit cession of original Indian title to the federal government, not directly to settlers. As Baring wrote of Knox and the Maine realtors to his fellow bankers, "The holders do not pretend to possess these titles and avow that the lands must be purchased of

the Indians."[12] It's curious, then, that the Commonwealth sidestepped the Congressional ratification process mandated by the Non-Intercourse Act each time that it treated with the Wabanaki Confederacy.

Treaties reflect the sovereignty of governments. For tribal nations, these formal agreements signified that they held authority over a defined territory. Some have argued that the Commonwealth of Massachusetts assumed exemption from the requirement for Congressional ratification because Massachusetts was one of the original Thirteen Colonies. But this inference of power proved porous. President George Washington, upon signing the Non-Intercourse Act, spoke to the Seneca Tribe of New York to explain, quite plainly, the applicability of the law. "No State, nor person, can purchase your tribal lands, unless at some public treaty, held under the authority of the United States," he said.[13]

Upon statehood, Maine not only adopted the Commonwealth's unratified Wabanaki treaties, but also the arrogance of ignoring these declared rules of colonization. A "timber rush" was unleashing all along the Penobscot River, and the dispossession of Penobscot lands took on a more aggressive approach—one officially sanctioned by the state government. Almost immediately, laws were enacted to control tribal affairs as a way to prosper from the river valley. Whether the Wabanaki tribal nations knew their sovereign rights were being violated is unclear from the dozens of petitions Natives filed with Maine.[14] The documents reveal a wave of grievances that Indigenously defined the statehood era: continual land loss, corrupt Indian agents, mismanaged tribal accounts, and, for the first time, chronic in-fighting.

In 1833, the Penobscots lost their last four townships secured to them in their 1818 treaty.

All they had left were a smattering of islands, including the main Indian Island—together accounting for less than one percent of their original homelands.[15] Maine deeded the territory to itself and reserved a $50,000 bond for the Penobscots. Outraged, tribal members marched some ninety miles in the dead of winter to complain at the state capitol. The sale was fraudulent, they told lawmakers, claiming that Indian agents had filed false signatures from false "chiefs" to approve the title exchange. The ordeal sowed confusion and undue division within the tribe. But the Penobscots lacked the political influence to keep the governor from finalizing the sale.

A decade or so later, the tribe complained about the matter again, this time over how the state was spending their $50,000 bond. The Penobscot wanted control of the funding but the Standing Committee on Indian Affairs told the tribe it didn't trust their tribal leaders. "If on the other hand the [Penobscot] Governors get possession of this money there is a danger," wrote the Committee. "It might be appropriated to the reward of particular services while the poor and indigent were left to suffer."[16] And just like that, without their consent, the tribes of Maine had become wards of the state, subject to a forced and oppressive culture of dependency.

In America's nascent stages of colonization, Natives were rarely aware of the tenuous nature of their sovereignty. Even by the 1960s, as the Wabanaki nations became more informed of their Indigenous rights, defending them was difficult. Few if any legal advocates at the time specialized in the complex field of federal Indian law. In Maine, the bigger obstacle was responding to plain racism. No attorney would represent Native Americans until Don Gellers arrived. While he didn't know it, the young

Jewish attorney from the Bronx wound up laying the groundwork for the Maine Indian Claims Settlement Act (MICSA). In the process, he also discovered that state officials were looting from the tribe's trust funds.

Gellers would pay a tall price for his advocacy: In 1968, he was framed for marijuana possession organized by Maine's Attorney General's Office.[17] The charges ended his Indigenous rights work and rendered him a fugitive. "Many have long claimed that a motivation to arrest Mr. Gellers was not just to enforce the state's criminal laws, but also to thwart his outspoken political and legal advocacy," said Mills the day she pardoned Gellers posthumously in January 2020.[18]

Through interviews, court records, Indian agent correspondences, and treaties, I analyzed more than a dozen historic cases centered on Indigenous land loss—the loss of territory that was addressed in some of the first federal laws enacted in the US. The Constitution acknowledges treaties as the "supreme Law of the Land." For the Penobscot Nation, the colonial continuum is especially acute. Indian Island, the tribe's main village, only became accessible by car in 1950. Despite such traces of modernity, the Penobscot's riverine culture and subsistence economy remained intact through the 1950s. But by the end of the decade, the river had become polluted with hazardous sewage, paper mill sludge, and industrial trash. This hampered fishing in the region, including of the Penobscot's famed Atlantic salmon, which today is an endangered species. Historians, tribal members, and state officials told me that for all the litigation over languishing land disputes, for all the division that has been sown, you'd be hard-pressed to find anyone who disagrees about the importance of restoring the vitality of the Penobscot River.

Fewer accept the virtues of the centuries-old treaties.

~

Indian Island is humble, like many reservations, with winding roads and simple subdivisions lined with nearly identical HUD homes. When I arrived in fall 2019, I bided my time waiting for Chief Francis by strolling the tribal cemetery, a wooded lawn whose markers evoke a story of Wabanaki kinship among the Sockalexises, Danas, Nicholases, and Rancos. It's also where Francis laid to rest his father two years ago.

Earlier that day, I met with Penobscot Nation Tribal Ambassador Maulian Dana at tribal headquarters, where a child's painting of the Penobscot River hung in the hallway down from her office. At thirty-six, Maulian is the Penobscot Nation's first appointed ambassador. Depending on which side of the political divide, she said Mainers either call her Princess Runs-Her-Mouth or Ms. Powerhouse. The daughter of the tribal chief who preceded Francis, Barry Dana, she is today's face of the Penobscot Nation. Around lunchtime, we met in an airy conference room and talked about her storied connection with the Penobscot River.

In July 1980, on the eve of the MICSA, Maulian's grandmother, Lorraine, testified before a US Senate Committee, telling lawmakers how her son, Barry, was raised to live off the land, tracking deer and catching fish to support the family. A single mother of five, Lorraine was unemployed when she traveled to Washington, DC to address the controversial land dispute. She worried that any outcome of the settlement could result in the loss of her tribe's two large islands. More than anything, she feared state control over Penobscot territory, a threat to her family's survival. "My son hunts and fishes my islands," Lorraine told lawmakers. "My family will endure hardship because of the control of the taking of deer and fish."

The islands were not lost after the MICSA, and Barry continued to forage the Penobscot. On weekends from attending the University of Maine, he found himself reconnecting to ancient ways upriver, where his grandfather had lived. It became spiritual for him. Drawn to Penobscot oral history, he developed a bond with *Ktàtən* or Katahdin, the tribe's most sacred mountain. He began running a hundred miles to its peak. Today, the journey is an annual Penobscot tradition.

Maulian, thirty-six, was seemingly groomed to be an Indigenous diplomat. She attended high school off the reservation in nearby Bangor, where she became incensed over Native American-themed mascots. In 2000, she witnessed her father, newly elected as tribal chief, take a stand for tribal sovereignty against the

state of Maine. The dispute centered around what may have seemed to some like an arcane issue—wastewater discharge—but to the Penobscots, it addressed what had always lived at the core of their colonization: jurisdiction of the Penobscot River. Two decades had passed since the MICSA, and Maine's paper mills had polluted the river to the point of serious health concerns. Chief Dana argued the pollution violated the tribe's right (as spelled out in the MICSA) to hunt and fish along their namesake river on their reservation, which the tribe maintains encompasses Indian Island and a sixty-mile stretch of the Penobscot River, bank

↑ Jenni Monet   *Lincoln Pulp and Paper Mill (formerly)*

Jenni Monet *Maulian Dana, Penobscot Nation's first appointed ambassador*

to bank. The Passamaquoddy Tribe joined the fight. Together, they petitioned the Environmental Protection Agency (EPA) to defend their treaty-protected riverine rights as Maine lobbied the federal agency to claim sole control of the state's watersheds.

In an unlikely scenario, Chief Dana and two Passamaquoddy leaders found themselves up against Maine's Freedom of Access Act. A state superior judge found the men in contempt of court for failing to turn over their EPA correspondences when three powerful paper companies formally requested the documents. The chief and the others were shockingly sentenced to jail; they each faced a thousand-dollar fine for every day they had ignored the court's order. The companies' use of the Act was a sneaky ploy to keep polluting, Chief Dana told the press, by diminishing the tribes' treaty rights.

"It was really a back door approach to get the tribe to be nothing more than a municipality," he said, referencing a passage of the MICSA.

Under the 1980 agreement, the Wabanaki nations fell under an odd provision that applied Maine's municipal laws to the tribes except in cases involving internal tribal matters. While the MICSA was intended to clear up muddled titles, it ushered in new confusion: where do states' rights end, and where does tribal sovereignty begin? A lawyer representing the Penobscot Nation, Kaighn Smith, Jr., acknowledged the imperfect outcome of the act—a jurisdictional mess that left ambiguous what defines "internal tribal matters" free from state regulation. "The tribes very much wish they could move to be who they are as they have understood themselves to be for many centuries," Smith told the *Washington*

Post. "They are not municipalities of the state of Maine."[19]

Maulian, then a high-schooler, was, like many tribal members, puzzled by the idea that state laws could govern the sensitive communications of tribal nations, governments that pre-existed the US. Their ancestors had lived their entire lives along the river. "You have it in your head that things are going to be fair when you go into a fancy court," said Maulian, who, at the age of fifteen, watched the judge order her father's arrest. "I was surprised."

For the 2001 Freedom of Access Act hearing, the paper companies—Great Northern Paper, Georgia Pacific, and Champion International—hired Pierce Atwood, a firm with a long history of litigating on behalf of the industry, as their representation. In arguing the case, their attorney stirred up rhetoric that had mired the MICSA controversy for decades, claiming that the settlement never intended to maintain the tribes of Maine as "nations within a nation." A deal's a deal, he argued: the tribes had entered into an agreement that subjected them to "limitations of a municipality."

"They are trying to undo what they did in 1980," said Pierce Atwood's Matt Manahan, as the tribes worked to appeal the court order. "Since they can't do it legislatively, they are trying to do it judicially."[20]

Eventually the tribe gave in to the paper companies, but not without protest. When Chief Dana announced on the courthouse steps that he would appeal his arrest to protect the tribe's documents—and the Penobscot River—he held up a plastic bottle of yellowish water that he said he'd taken from the river. He dared paper mill executives and their attorneys to drink from it. The provocation was emblematic of the contentious history between the Penobscots and the mills. After all, it had been

the tribe's water monitoring program that led to hefty fines for the companies months before Dana and the others were sentenced to jail. A state inspection of Great Northern, Georgia Pacific, and Champion International uncovered a litany of water quality violations: illegal wastewater discharges, falsifying water quality records, operating without permits.[21] One mill was found to have skirted mandatory water testing for nearly a decade; state regulators noted the cobwebs they found in the company's sampling equipment.[22]

From 2000 to 2002, the battle to control Maine's waterways bounced from federal to state court, and, for a moment, the Supreme Court, which refused to hear an appeal from the tribes. There was even an attempt to resolve the dispute with Maine's then-governor, Angus King, but those efforts quickly died. Realizing all legal options had been exhausted, Chief Dana and his Passamaquoddy colleagues accepted defeat; they'd turn in the documents while making a statement about the injustice they felt. Before dawn one morning in May 2002, the men bowed their heads and prayed in the old way, on somber ground, the site of a centuries-old Wabanaki massacre. Then they set out on foot for the state capitol, walking for thirty-three miles with the blood memory of their ancestors—a presence that ran through them like holy medicine.

～

Nearly twenty years passed between that grave morning and the day Chief Francis met me in the spacious conference room where Maulian and I had spoken. Like every tribal leader I've known, the Penobscot chief is constantly on the go. In late 2019, he and Chavaree had yet to resolve their latest river battle. The leaders had also recently joined the state task force focused on renegotiating the terms of the MICSA to restore tribal sovereignty. In the talks, Francis worked to explain to lawmakers how autonomy should function for the Penobscot Nation— free from state control. To bolster this education, Chavaree brought in an outside legal expert to introduce the basics of Federal Indian Law, highlighting the fact that in the US there are three forms of recognized governments: federal, state, and tribal.

The tribe's latest battle coincides with the Penobscot Nation's discovery of new allies in a coalition of river protectors that banded together to revive the depleting sea-fish run on the Penobscot River. In 2012 and 2013, the Penobscot River Restoration Project led to the removal of two of Maine's oldest dams. For the Penobscot Nation it was a dream come true. "I never thought in my lifetime that I would see dams being removed from the river," said tribal elder Butch Phillips, beaming in one of the project's campaign videos. "The river is free-flowing here, as my ancestors experienced it."[23]

As the first of the two dams were being dismantled in August 2012, the Penobscot Nation received a letter from Maine's attorney general, stating the Penobscot "River itself is not part of the Penobscot Nation's Reservation."[24] Chavaree and Francis were stunned. The tribal leaders had no choice but to take legal action. The state maintains that the tribe's reservation boundaries are limited to islands only—not the river or the riverbed that connects to the Nation's tribal lands.

The issue over who has jurisdictional rights to the sixty-mile main stem of the Penobscot River was a sore point. Four years before that letter was received, state game wardens complained about tribal wardens enforcing Penobscot Nation permitting laws on non-Native hunters. For its part, the tribe had accused state officials of similar offenses, only on tribal

citizens utilizing the river on their own tribal lands.

The Penobscot Nation ultimately lost their case against Maine in federal district court in 2015, and then again in 2017 when they appealed to the First Circuit. Then-governor of Maine, Paul LePage, opposed the Penobscot Nation's reservation challenge and the Penobscot River Restoration Project, specifically the dismantling of any dams. But the Department of the Interior had intervened in the case, arguing on the tribe's behalf: "If the Penobscot River in its entirety is off-reservation, it is not clear where the Nation can meaningfully exercise this federal statutory fishing right." And while it seemed this loss in their riverine rights fight would be impossible to overturn, Judge Juan R. Torruella, in his dissenting opinion, offered a glimmer of hope.

Criticizing the majority's ruling as another example of breaking treaties with Native Americans, he referenced Lorraine's locution, "fish my islands," as a modern indication of the Penobscot's treaty-protected fishing rights. "According to [Maine's] interpretation, it would appear that the Penobscot Indian Reservation shrinks when the water levels in the River rise, and then expands when those levels fall," wrote Torruella.[25]

The tribe had one other legal option. It could ask the First Circuit for en banc review, a rare rehearing of the case before the court's entire bench of active judges. After three long years of waiting, the Penobscot Nation learned in March 2020 that their request had been granted. Oral arguments were heard in September. Helping the state argue their case was Pierce Atwood, the same firm that represented Maine's paper companies when the court sentenced Chief Dana to jail. A decision could come by the end of the year.

→ **Jenni Monet** *Penobscot River*

~

"[The] technical definition leaves no room for surrounding waters," Attorney General Aaron Frey wrote in response to the tribe's 2019 petition for rehearing.[26] Frey, who would argue against the tribe in 2020, relied on dictionaries to support the state's position. "Island," in its plainest reference is "a piece of land that is completely surrounded by water," according to the *Oxford English Dictionary*, or "a tract of land surrounded by water and smaller than a continent," as sourced from *Merriam-Webster*. In 2017, these sources had been enough to appease an appellate court ruling that favored Maine. On the eve of Maine's bicentennial, the state remained defiant.

In June 2019, Governor Janet Mills signed into law a bill that protects sustenance fishing for the tribes of Maine, establishing the first water quality criteria aimed to specifically uphold traditional Wabanaki fishing practices. The governor touted the legislation as one that promised tribes "the most protective fish consumption rate in the country," and Frey said there was no contradiction between this law and the state's position in the lawsuit. But the way tribal leaders see it, if the Penobscot Reservation does not include the river, the sustenance fishing law is merely symbolic.

A week later, Francis and Chavaree held a special meeting on Indian Island about whether to make the Penobscot River a formal tribal citizen. Francis said that it was the next best step in asserting the tribe's treaty rights. Chavaree agreed. "Penobscot sovereignty is our ability to live our lives consistent with our worldview, as we have for thousands of years," he said. As the meeting emptied out, Maulian emerged from the small crowd. Roughly two dozen votes had been cast to adopt the Penobscot River as a tribal citizen. Following the White Earth Band of Ojibwe's similar granting of rights to wild rice in Minnesota, the Penobscot Nation made quiet history the day they became the first tribe in America to formally adopt its river.[27] But there were still more details to iron out. "I just hope people understand, in time, what a great thing we did," she said.

In the year since, tribal leaders have worked on beefing up their legal team in preparation for their First Circuit rehearing. The Penobscot Nation, stymied under state control, isn't wealthy, but their humble means haven't deterred them from addressing many of the same issues that concerned the Penobscot centuries ago. "Today, the methods are much different, but the issues and what we want remains the same," said Francis.

Chavaree told me that he hasn't given up for his tribe, his family, or himself. But in his quiet nature, a weariness showed. "We definitely have had our share of disappointments over the years, but it's good that we still have a spark of hope—that we're not so far gone that we don't feel any sense of encouragement looking ahead," he said.

By early August, Francis, Chavaree, and Maulian Dana tuned in on Zoom to what would be the last meeting of the months-long MICSA state task force. As many as twenty people squeezed together on the screen in their individual video boxes, from lawmakers to attorneys to tribal leaders. Francis made a few lasting remarks. "The tribes of Maine aren't interested in being perpetual victims," he said. "You're not personally responsible for how we got here or for the historical tragedies that have taken place with Maine's tribes." Maulian, situated in the video box next to the chief, listened intently. Francis continued. "It's our hope that this conversation results in real change, for real people, for a real condition," he said, and called for a future that promised trust between the tribes and the state of Maine. It was more of a wish list than a plan—an expressed fantasy that, for the Penobscot Nation, their sense of injustice could be overturned by the decisions of their colonizers.

The tribal leaders had no idea when the state legislature would resume their annual session. The coronavirus had stalled everything, as everywhere. The twenty-two task force recommendations[28] would be tabled as lawmakers squabbled over whether to call for a special session, which the governor seemed reluctant to do.

A few days after the meeting, I turned to Maulian's Facebook page, where none of her loyal followers had cared to comment about the task force. Instead, her post about the removal of a controversial statue in a neighboring town had triggered heated dialogue. A towering figurine of a fictitious Indian chief, Hiawatha (not to be confused with the authentic Onondaga chief), had been ordered to be dismantled. Maulian, a longtime activist against racist mascots, remained silent as her frustrated audience responded to Mainers who saw the statue as somehow respectful to the Wabanaki nations. "Indigenous people have made clear that this shit isn't respectful," read one ally. "They're still here and this is still their home." ○

Notes

1. Maggie McKeon, "Penobscot Nation and their River," Colorado College, November 18, 2013.

2. Harold E.L. Prins, "The Penobscot Nation's Reservation of the Penobscot River Accompanying Its Reservation Island in the Penobscot River in the 1796 and 1818 Treaties with Massachusetts and 1818 Treaty with Maine," Prepared for the Penobscot Nation in *Penobscot Nation v. Janet T. Mills*, et al., Civil Action No.1:12-cv-00254-GZS, December 11, 2013, 80.

3. Prins, 79.

4. Prins, 80.

5. Reflecting on the treaty-making process in more general terms, Judge Lorenzo Sabine, an attorney born and raised in Maine, found that "there is record of dishonest and ignorant interpreters at the 'talks' or conferences; of incompetent and ill-disposed commissioners, who stated their terms in vague language, or disposed of the business with which they were entrusted in hot haste, and before the chiefs could understand what was required of them." See Lorenzo Sabine, "Indian Tribes of New England," *Christian Examiner and Religious Miscellany*, vol 27 (March 1857): 27–55.

Sherman F. Denton *The Atlantic Salmon*

6. Smithsonian Insider, "The 'Indian Problem,'" YouTube, May 26, 2016.
7. Richard Peters, ed., "An act to regulate trade and intercourse with the Indian tribes, July 22, 1790," *The Public Statutes at Large of the United States of America* (Boston: Charles C Little and James Brown, 1845), 137–138.
8. Native American Rights Fund, "The Eastern Indian Land Claims," *NARF Legal Review* 7, no. 1 (October 10, 1980): 1.
9. See "Chapter IX: Problems of Speculation" and "Chapter X: The Sale to Baring," *Colonial Society of Massachusetts*, Volume 36, *William Bingham's Maine Lands 1790–1820* (Boston: Colonial Society of Massachusetts, 1954).
10. Alexander Baring to Hope & Co, May 26, 1796, in *William Bingham's Maine Lands 1790–1820*, Colonial Society of Massachusetts, 653.
11. Prins, 74–75.
12. Baring to John Williams Hope, December 8, 1795, in *William Bingham's Maine Lands 1790–1820*, Colonial Society of Massachusetts, 653.
13. "From George Washington to the Seneca Chiefs, 29 December 1790," *Founders Online*, National Archives.
14. Maine Department of Health and Welfare and Margaret Snow, "Indian Affairs Documents From Maine Executive Council: 1830–1839," May 17, 1935.
15. According to *Maine: An Encyclopedia*, the Penobscot reservation was 4,866 acres as of 2007. This represents a loss of approximately ten million acres of land. See Samantha Carrs Pierce, "Penobscot Nation v. Janet Mills: How Two Sides Understand One River," (bachelor's thesis, Bates College, Lewiston, Maine, 2019), 186.
16. Maine Department of Health and Welfare and Margaret Snow, "Indian Affairs Documents From Maine Executive Council," no. 2, (1939): 199.
17. Colin Woodard, "Gov. Mills grants full pardon to late tribal attorney Donald Gellers," *Portland Press Herald*, January 7, 2020.
18. State of Maine, Office of Governor Janet T. Mills, "Governor Mills Issues Posthumous Pardon for Former Passamaquoddy Advocate Donald Gellers," January 7, 2020.
19. Pamela Ferdin, "Maine Court Raises The Stakes in Fight Over Tribal Rights," *Washington Post*, November 11, 2000.
20. Murray Carpenter, "Sovereignty in Jeopardy: the Great Paper Chase," *Portland Phoenix*, November 9, 2001.
21. Carpenter.
22. Carpenter.
23. The Nature Conservancy, "Butch Phillips: Restoring the Penobscot River," YouTube, May 11, 2010.
24. Maine Attorney General William J Schneider to Commissioner Chandler Woodcock and Colonel Joel T. Wilkinson, August 8, 2012.
25. *Penobscot Nation; United States, on its own behalf, and for the benefit of the Penobscot Nation, v. Janet T. Mills et al.,* 16–1424, (1st Cir. 2017).
26. *Penobscot Nation; United States, on its own behalf, and for the benefit of the Penobscot Nation, v Aaron M. Frey et al.,* 16–1424, (En banc 1st Cir. 2019).
27. Winona LaDuke, "The White Earth Band of Ojibwe Legally Recognized the Rights of Wild Rice," *Yes Magazine*, February 1, 2019.
28. Caitlin Andrews, "Maine panel issues milestone report on tribal relations with gambling looming as main hurdle," *Bangor Daily News*, January 14, 2020.

John Trudell

LOOK AT US

Look at us
We are of earth and water
Look at them
It is the same
Look at us
We are suffering all these years
Look at them
They are connected
Look at us
We are in pain
Look at them
Surprised at our anger
Look at us
We are struggling to survive
Look at them
Expecting sorrow be benign

Look at us
We are the ones called pagan
Look at them
On their arrival
Look at us
We are called subversive
Look at them
Descending from name callers
Look at us
We wept sadly in the long dark
Look at them
Hiding in technologic light
Look at us
We buried the generations
Look at them
Inventing the body count
Look at us
We are older than America
Look at them
Chasing a fountain of youth

Look at us we are embracing earth
Look at them
Clutching today
Look at us
We are living in the generations
Look at them
Existing in jobs and debt
Look at us
We have escaped many times
Look at them
They cannot remember
Look at us
We are healing
Look at them
Their medicine is patented
Look at us
We are trying
Look at them
What are they doing
Look at us
We are children of earth
Look at them
Who are they

Karen Washington

THIS VIRUS WILL NOT DEFEAT ME

What is this disease called COVID-19, coming into my community and taking lives? Before this stranger came, so many of us were already suffering. Our food system is tainted with additives, causing diet-related illnesses. Our water supply is laced with chemicals, so we can't drink from our own kitchen sinks. Our air is polluted, so we can't breathe in our own streets. Then this stranger came knocking on our door, taking away family, friends, and kinfolk. We mourned alone in silence, unable to touch, hug, or kiss our loved ones goodbye. The coldness of solitude is deafening.

What is this disease called COVID-19? Why can't this virus feel my heartache, the lives lost, the jobs lost? Cities that were once vibrant turned into ghost towns. We walk around as strangers with swaths of fabric tied around our noses and mouths, trying to breathe a new normal.

You hear folks ask, when will this virus go away, or will it come back?

My question is, when will my people find justice? The hardest hit communities have been the poor and people of color. We are dropping like flies, yet the country wants to spring its shop doors open as if Black Lives don't matter. As the United States approached over one hundred thousand lives lost, too many failed to take a moment to remember that number has meaning. Each one of those one hundred thousand was a human life—a mother, a father, an aunt or uncle, a child.

They began to call us essentials. But they did not notice we were essential until COVID-19.

Who remembered the endless hours, days, weekends, and holidays we'd sacrificed to keep this country afloat? As the pandemic wore on, we stood in long lines for food. We stood with humility, confusion, and despair setting in as our bills piled up, our mortgages and rents unpaid, our account balances at zero.

My country 'tis of thee, sweet land of liberty, of *thee* I say! COVID-19 has pulled back the cover to show the inequities that underlie the true America—a divided America. The Have and Have Not America. So what are we going to do about it? Leadership at the top has failed us! We cannot go back! Things must change! Things *are* changing. We must struggle through COVID-19. We must work with a vengeance to do better as a society.

This virus teaches us two things: you can't hide, and it can kill you—no matter who you are. We all feel the effects of the virus. Staying home brings many of us closer together, the days run into each other as though time were a single unbroken continuum, and the question everyone asks is, "Who's Zoom-ing who?"

I can only speak for myself for what I have done and continue to do. I have attacked COVID-19 by growing food, growing community. Loving more, caring more, sharing more. I continue to appreciate the clear sky, the smell of rain, the silence at night. I continue to imagine how it felt with fewer cars on the road. The air smells and feels different. I sit and ponder as birds sing and squirrels scamper along the tree trunks. Neither seem to mind me. There's now time for reflection and time for promise.

Jenevieve *Root Harvest*

You see, I will not let this virus defeat me, take away my power to make change or to be a change maker. My hope is for a kinder America, a compassionate America, a forgiving America. A loving America.

But in the meantime, I will remember this virus. I will respect the aftermath of this virus the best way I know how—by continuing to listen to science and truth rather than false-hoods and speculations. I am going to heed the warnings so that I can protect myself and the children, now and for future generations. I am a better person because of this virus and I am up for the challenge to defeat future pandemics the best way I know how, with bravery and common sense.

Stay safe, and be strong. ○

Gerrard Winstanley

THE DIGGERS' SONG

You noble Diggers all, stand up now, stand up now,
You noble Diggers all, stand up now;
The waste land to maintain, seeing Cavaliers by name
Your digging do disdain, and persons all defame.
Stand up now, stand up now.

Your houses they pull down, stand up now, stand up now,
Your houses they pull down, stand up now;
Your houses they pull down to fright poor men in town,
But the Gentry must come down, and the poor shall wear the crown.
Stand up now, Diggers all!

With spades and hoes and plowes, stand up now, stand up now,
With spades and hoes and plowes, stand up now;
Your freedom to uphold, seeing Cavaliers are bold
To kill you if they could, and rights from you withhold.
Stand up now, Diggers all!

Their self-will is their law, stand up now, stand up now,
Their self-will is their law, stand up now;
Since tyranny came in, they count it now no sin
To make a gaol a gin, to starve poor men therein.
Stand up now, stand up now.

The Gentry are all round, stand up now, stand up now,
The Gentry are all round, stand up now;
The Gentry are all round, on each side they are found,
Their wisdom's so profound to cheat us of our ground.
Stand up now, stand up now.

The Lawyers they conjoin, stand up now, stand up now,
The Lawyers they conjoin, stand up now;
To arrest you they advise, such fury they devise,
The devil in them lies, and hath blinded both their eyes.
Stand up now, stand up now.

The Clergy they come in, stand up now, stand up now,
The Clergy they come in, stand up now;
The Clergy they come in, and say it is a sin
That we should now begin our freedom for to win.
Stand up now, Diggers all!

The tithes they yet will have, stand up now, stand up now,
The tithes they yet will have, stand up now;
The tithes they yet will have, and Lawyers their fees crave,
And this they say is brave, to make the poor their slave.
Stand up now, Diggers all!

'Gainst Lawyers and 'gainst Priests, stand up now, stand up now,
'Gainst Lawyers and 'gainst Priests, stand up now;
For tyrants they are both, even flat against their oath,
To grant us they are loath, free meat and drink and cloth.
Stand up now, Diggers all!

The club is all their law, stand up now, stand up now,
The club is all their law, stand up now;
The club is all their law, to keep poor men in awe;
But they no vision saw to maintain such a law.
Stand up now, Diggers all!

The Cavaliers are foes, stand up now, stand up now,
The Cavaliers are foes, stand up now;
The Cavaliers are foes, themselves to disclose
By verses, not in prose, to please the singing boys.
Stand up now, Diggers all!

To conquer them by love, come in now, come in now,
To conquer them by love, come in now;
To conquer them by love, as it does you behove,
For He is King above, no Power is like to Love.
Glory here, Diggers all!

Note

An advocate of the commons and critic of private property, Gerrard Winstanley led the English Diggers in the 1649 occupation of St George's Hill, where they proceeded to plant parsnips. "True freedom," Winstanley wrote, "lies where a man receives his nourishment and preservation and that is in the use of the earth."

Lucy Zwigard

A FARMERS MARKET IN THE GLOBAL PANDEMIC

Reflections from Berkeley, California during COVID-19

It's a sunny, calm morning in downtown Berkeley as vendors pop open shade canopies and set up freshly furnished handwashing stations, preparing to deliver nutrient-dense food to urban streets. At promptly ten o'clock, the market managers open designated entry points for the shoppers lined up at either end of Center Street, the marketplace corridor bordering Martin Luther King Jr. Civic Center Park. I wrangle on my plastic gloves and approach the tower of warm pastries from Frog Hollow Farm, the stone fruit orchard where I work. In the time of COVID-19, shoppers can't seem to resist our flaky, buttery cherry turnovers or Meyer lemon tarts. Something about being stuck at home fuels a societal sugar addiction. My coworker and I don't hold back, suggesting a scoop of vanilla ice cream to top the apple blueberry crisp.

Because they are crucial to local economies embedded in a year-round growing season, the markets were deemed essential by the state of California. At the Downtown Berkeley Farmers' Market during the first week of May, we sold every single pastry item that we brought—as well as every box of Royal Tioga cherries, the season's first.

A woman who purchased our apricot conserve to pair with her homemade cornbread told me that she had baked cookies for the first time in ten years. Forced time at home has fostered an increase in DIY home, garden, and kitchen projects—what one might call endeavors in "urban homesteading." Customers ask more questions than ever before about shelf life, preservation, fermentation, and how to freeze food for later use. With restaurants closed, more are cooking meals from scratch and realizing the cost effectiveness and nutritional benefits therein. Fear of infection and illness paired with recurring reports of COVID-19's rampant spread through industrial food facilities have piqued the public's interest in high quality, fresh food. Consumers who know their farmer may better recognize the superiority of produce sold at local markets, like ours in Berkeley. I hope that the wave of support for local farms will outlast these pandemic times.

Vendors in the Bay Area have seen an increase in market revenues ever since the first "shelter-in-place" mandate went into effect mid-March. Local bakers are selling out of breads; bulk sales of shelf-stable foods like honey, olive oil, and jams are up; Frog Hollow even started selling frozen pastries as "take and bake" items. CSA participation has drastically increased, helping cover sales lost due to the closure of restaurants and other wholesale buyers. (However, the cost of packaging and

Jo Zimny *Jacob Eisman, Six Circles Farm in Upstate NY*

the labor involved in smaller-scale, more diversified produce boxes is significant.) Farmers markets and their smaller, less complicated supply chains are more agile than grocery stores and can respond quickly to these present needs. Produce vendors, like the wonderful folks at Riverdog Farm, now pack veggies into bags during their market set up (as opposed to selling out of crates) to more easily facilitate the socially-distanced shopping experience. As new guidelines crop up, resiliency in the mar-

ketplace depends on working together to plan creatively, communicate effectively, and share resources with partners and patrons.

The market managers and staff have worked tirelessly to maintain a smooth shopping experience and a sense of security amid the new regulations. The layout of the market has shifted to accommodate safety protocols. Vendors that draw major crowds are separated from the primary market area to allow for more space. One of the most difficult outcomes is lines, which are necessary for entering the market itself as well as shopping at each booth. If people normally dislike waiting in line, they especially hate it when being away from their house is said to be a public health threat. Weary side glances and judgmental shifts in body language are commonplace among shoppers navigating the six-feet rule, as are exasperated bouts of laughter to make light of the absurdity of human nature in anxiety-driven times.

The markets serve as an important space for social interaction while we shelter in place. Many folks purchase items as gifts for neighbors who are quarantined in their homes; others come with lists for their friends and family who might be immunocompromised or otherwise avoiding public settings. I ask myself how the markets maintain a sense of normalcy in these shifting times, and simultaneously contribute to the creation of new norms. One woman waiting in line for my booth was clearly disheartened by the new systems in place, and exclaimed with a sigh, "I feel like I am visiting another planet!"

Technological tools like card readers come in handy to minimize contact between customer and vendor—especially when people can pay by simply hovering their smartwatch or other device over top. While the markets are intended for everyone, including customers experiencing poverty, the use of these integrated technologies serves a privileged demographic. Thus, the marketplace is yet another arena where safety measures and responses to COVID-19 intensify class disparities, as those of us with greater access to capital and infrastructure stock up on food and other supplies.

The fear of scarcity that pervades American culture is coupled with denial of our dependence on massive-scale production schemes and ignorance of the ways in which the systematic abundance of industrial food breeds widespread shortages of nutrient-dense food. The capitalist food regime is faltering in the face of COVID-19 as effects of the virus underscore the imperative for establishing community food sovereignty. In what ways will this serve as a wake-up call to galvanize the movement for regional, regenerative food systems?

For me, sheltering in place has clarified the importance of farmers markets in the Bay Area and beyond. The markets are a keystone species in the food supply ecosystem, one that supports responsible land stewardship and quenches the palate for robust food. Not only our bodies but our souls are fed when we visit the farmers market; they provide us with a direct link to the land, a sense of freedom, healing sustenance, and the inevitability of life and death, from which American culture shelters us. Those of us who crave heartfelt connection, groundedness, and land-based spirituality can find this amid hardworking farmers, their gleaming smiles and calloused hands not easily disguised by masks or gloves. These land stewards provide much needed strength and solace to city dwellers in this existentially challenging period. ○

Ben Prostine

THE PREAMBLE TO THE INTERSPECIES LIBERATION MOVEMENT OF THE WORLD

On May 31, 2019, the 200th birthday of Walt Whitman, one man and a guerilla band of cattle gathered in the savanna pastures of Wisconsin to conspire. Among the trees and grass, and outside the reach of antenna waves, they formed the Interspecies Liberation Movement of the World (ILMW). After much rumination, the following preamble was drafted:

Noting our kinship with the animals, humans, soils, and plants of the world,
We make a declaration of inter-dependence.

In solidarity with our comrades,
Those confined in CAFOs, sweatshops, war-zones
And other infernos of the global economy,
We make a vow: To labor for the liberation
Of all live-stock confined by capital.
No more shall the rich sit on
the hides of our brethren.

And we make a pledge:

We pledge allegiance to the grass
And to the ecosystem for which it stands,
One earth, under the sun, with no despots,
But with peaceful rumination for all.

Today we graze the grasslands of Wisconsin,
Tomorrow the lawns of the White House.

A movement was made to adjourn and the herd proceeded to graze, but not before a reading from Whitman's *Leaves of Grass*, which reminded them that the American project may call for "the ceaseless need of revolutions," that their duty is to "horrify despots," that the mantra of democracy is "whoever degrades another degrades me," and the great question of literature is still: "what is the grass?" ○

Patriotism, like other virtues, is easily counterfeited. Gruff old Dr. Johnson called it "the last refuge of a scoundrel." It has one thing in common with charity, "it covers a multitude of sins." It often expends itself in mere bawling. Our holiday oratory brings out no end of inspired and inspiring utterances; but allowance ought to be made for considerable leakage of gas. Indiscriminate praise of everything American is a cheap way of drawing applause, but the truest friends of the country are they who make us worthier to be free, who

  help to save mankind,
Till public wrong be crumpled into dust,
And drill the raw world for the march of mind,
Till crowds at length be sane and crowns be just.

"Patriotism," *Tiller and Toiler*, July 28, 1899

↑ **Paola de la Calle** *Untitled (Earth and Resistance)*
→ **Alexis Hubert Jaillot** *Nova orbis tabula, ad usum serenissimi Burgundiae Ducis*

Mapping •
Networks

County Planning Commissions, Extension Programs

Charlotte Du Cann

BEANS, PEAS, QUINOA, WHEAT

"OK, now you need to sort them," I say, pouring a stream of seeds onto the table.

"Oh, you mean for real?" asks one of the fifteen workshop participants.

I laugh. "Yes. There are six kinds and they need to be separated into piles."

There is silence in the room as a sea wind moves among the tropical trees in the university garden. It's March 2020 and we have been exploring the myth of Psyche and Eros, and how this Roman tale holds keys about the regeneration of ourselves and the earth. Sorting the seeds is the first of four tasks the goddess Venus sets for Psyche, a human girl, as a test of true love for her immortal son. In the myth, a sisterhood of ants helps Psyche sort the seeds: wheat, barley, lentils, chickpeas, bean, and poppy, six staples of the ancient world that still underpin the food stores and medicine chests of modern civilization. In the workshop, many

light fingers get to work, discussing the differences and their provenance.

"They are all from the UK," I tell them, "And these chickpeas are the first to be produced commercially in these islands."

"Is that because of climate change?" asks another student.

"Yes and no," I replied.

Once, we had stories that kept us close to the land and told us about a spiritual and physical negotiation between the wild world and the domesticated crops that fed and maintained us. Since civilizations broke those contracts in favor of control enforced by patriarchal religions, science, and industry, we are now bearing the consequences of a culture that has ignored the tasks set by the goddess of love and beauty. The lessons which once taught human beings how to keep in harmony with the living earth have been forgotten.

Can we recover those kinds of relationships? Can ancestral myths and teachings speak to us again?

Master of the Die and Antonio Salamanca
Venus ordering Psyche to sort a heap of grain, from the Fable of Psyche

The seeds on the table that day in Cornwall came from a pioneering pulse and grain enterprise in East Anglia called Hodmedod's, named after the Suffolk dialect word for a snail or hedgehog. The magical tendrils of the beans and peas that began the business in 2012 have wound about my own writing projects for over a decade now, the pulses working as a metaphor for restoration at festivals and gatherings, and as an essential ingredient in a low-carbon cookbook. I've been in conversation with Josiah Meldrum, one of Hodmedod's founders, about the relationship between crops, language, and imagination ever since our culinary paths crossed at a "Grow Local" food conference he organized with our grassroots Transition group, Sustainable Bungay, in late 2008. We co-edited a small local food magazine and mapped out a book called *Roots, Shoots and Seeds* about the shifting nature of community-based food systems. We interviewed pioneer scientist, Martin Wolfe, who formulated the YQ composite cross population wheat specifically grown to withstand climate change. We discovered how the beans and peas could work as cultural intervention in a time where few people could spot the difference between wheat and barley—even though human beings have been eating bread and drinking beer for millennia—and where discussions about sustainable food rarely venture beyond the homely fences of horticulture and animal husbandry to consider the vast arable territories that feed and fuel the cities.

In autumn 2019, Hodmedod's held an event called Harvest Home where their farmer partners shared their experiences and practices. I spoke about the wide communications and knowledge gap between people and the fields in which crops are grown. The Bean Store—Hodmedod's office, packing rooms, and warehouse—is surrounded by thousands of hectares of industrally grown commodity crops, including sugar beet, barley, and oilseed rape. The challenge of shifting to a sustainable agriculture seems on the surface as impossible as the tasks Psyche was originally set. What we need, I said, is a story that can grab our attention and pull us into a different relationship with the land.

Josiah and I met afterward to discuss what this storymaking may look like, and he said, "You need to identify the wedge which you can then push and expand from. Climate change and biodiversity loss are overwhelming and soul-destroying concepts, but if the idea is that you can do a small thing and make a difference, that can empower everyone to take part."

Hodmedod's first small wedge into a different narrative was the field, or fava, bean. Instead of working with the non-native beans people are used to eating and attempting to grow them in the United Kingdom's difficult, different climate and terrain, it chose to focus on this small, brown, seemingly unglamorous, Iron Age staple that was still being grown but as animal feed or for export to Egypt, where it is an everyday ingredient. The story is about a forgotten plant that was being reimagined. It has history, a fairytale, and a regenerative nature (both for human beings and for the soil in which it is grown). And it also makes great hummus and dal, tallying with the rise of earthy and spicy North African and Middle Eastern cuisines in the city cafés and restaurants.

"We are sometimes viewed as entrepreneurs," Josiah told me, "but we are really campaigners and activists, who came from the NGO world and happened upon a business model that allows us to do our campaigning and activism. We knew we were not going to change the world by just growing beans that were destined to go to Egypt. So we asked ourselves: How do we create these beans as intercrops, how do we find other varieties, get them into hospitals and schools, into the hands of food writers who can then inspire people to cook them at home? Most producers are product managers, but we do everything: crop development, agronomy, sales, marketing, advocacy."

The first harvest and its successful publicity campaign enabled the company to expand into other crops and speak with other farmers, backing their experiments and giving them an opportunity to think differently about their land, to grow other crops, or to convert to organic production. That guarantee also allowed farmers to take an interest in what they were growing in a way that producing commodity crops for agronomists and financial markets never could. Farmers who had once grown anonymous wheat for traders were now growing up to fifteen ancient and new varieties, experimenting with agroforestry, building connections not only with their new producers but also with small millers (and chefs, shopkeepers, and customers) and finding their names on the labels. They were no longer cogs in a machine, but valued growers and providers.

Crops such as quinoa, buckwheat, lentils, chia, and lupin seeds, which previously had been imported, could now be grown and produced nearer to home, bringing the story about low-carbon, less damaging agriculture closer to home kitchens. This kind of flexibility with scale has allowed producers to be nimble and imaginative, demonstrating a different kind of bio- and human-diverse farming, based not on monocultural demand and high yield but on healthy, life-giving relationships.

"We don't work in a linear masculine way," Josiah said during a conversation about the web of correspondences they have built up over the years. "We have always concentrated on the network of supply. How can we share information in multiple directions? How can the consumer engage with the farmer and start thinking more ecologically, knowing how the decisions we make end up in the landscape and the quality of the wider world?"

For the last eight years I've been co-editing a journal published by the Dark Mountain Project, responding to the converging social and ecological crises of our time. Our new collaboration, Writing in the Field, is an exploration of how the two organizations we direct and work for might bridge that gap: to see what kind of creative story might emerge when Dark Mountain writers of fiction, non-fiction, and poetry go through the field gates and meet some of Hodmedod's farmers in different parts of Britain.

Hodmedod's—like farmers, like writers— are in the long game: an engagement with life and work that requires commitment, vision, and staying power. You can only plant and harvest once a year and must gauge the risks of new ways of working. This kind of attention runs counter to a consumer culture that thrives on start-ups and just-in-time delivery. However, the slow nature and style of the Suffolk snail and hedgehog might well prove the "winner" in

Marion Post Wolcott *Picking beans near Homestead, Florida*

the end, as we race towards finding some kind of social and planetary sustainability before the world tips into collapse.

"When we started," Josiah said, "we wanted a revolution. Everyone wants a revolution! Then we realized, it's the incremental changes that affect the revolution. And then you realize you have had a revolution. You just didn't see it coming."

When COVID-19 hit in the UK in March 2020, the way people regarded their food sources reset overnight. There was a dramatic shift in the nation's 24/7, pick-up-something-at-the-supermarket, on-the-hoof attitude. Suddenly, everyone was aware of the fragility of industrial supply chains and the value of local small businesses and the people who work in them. They started to stock up their store cupboards and get into home baking. As the UK went into

lockdown during "the hungry gap" (the time between the last harvest of native winter crops and the start of spring season), Hodmedod's was overwhelmed by orders and their team was working double shifts. They had to close down their website for over a month.

"Can I give you a hand?" I asked when the deluge arrived. "I can bring lunch!"

For several Sundays following the workshop in Cornwall, I began sorting the seeds on a grand scale, working alongside Josiah and Mark at the Bean Store warehouse, shoveling tons of Emmer wheat, naked barley, quinoa, green lentils, yellow split peas, carlin peas, marrowfat peas, and, of course, fava beans, into pallets of five-kilo brown paper bags.

Sometimes the myths get real. ○

Addison Bowe

NARRATIVES

I look around and see only beauty.

The sun broken by Cottonwood branches.

The crop fields muddy and looking hopeful for spring.

I hear frogs, a bit too early, and the air is threateningly crisp.

I smell wood smoke and notice visiting birds.

And yet, amidst all this serenity and peace, I am longing for the place that I belong.

I am not sure I belong among the beauty.

Nikki Mokrzycki
Old Laces

Caroline Shenaz Hossein

MUTUAL AID

Localizing Economic Cooperation in the Global South, Rural Areas, and Among Black Minorities

Mutual aid groups are places where people collectively lend and save among each other, and they have existed long before banking to the poor was ever named.[1] Scholar on community economies theory J.K. Gibson-Graham has repeatedly demonstrated that community economies have existed in spite of market fundamentalism and that understanding the various ways people live collectively is vital to unraveling the dominant view that there is only one way to organize business and society.[2]

I had always known about pooling resources growing up in a Caribbean immigrant household in a white-dominated society, but we did collectivity in our bubble. It was not until I lived in the small rural community of Athiémè in Benin, West Africa as a US Peace Corps volunteer that I saw how mutual aid could dominate local life. Because farmers and rural businesspeople were excluded from formal banks, they had difficulty accessing financial goods and other inputs to run their farms. As a result, mutual aid was pervasive there.[3] Years later, as an academic, I found that exclusion from goods and services in the Caribbean and Canada also turned people toward what they know best: how to help one another through informal cooperative groups.

It was political economist Karl Polanyi who first made it known that everyday people carry out double-movements alongside an extremist economy, thus making clear that markets operate with diversity.[4] Polanyi's masterpiece depended on his work in Dahomey (now Benin), a place that emphasized social life over money-related transactions. Mutual aid is a habit that all peoples have engaged in at some point in history and these groups continue in our modern world.[5]

That said, any trends towards mutual aid in the United States and Canada in the time of a crisis should be understood not simply or principally as a return to earlier giving habits, but also as an echo of ongoing giving practices among many people in rural communities, excluded minorities, and those living in the Global South.[6]

It is human nature to gravitate towards groups we know and trust. Russian anarchist and philosopher Petr Kropotkin,[7] one of the first scholars to document mutual aid, detailed living creatures of eastern Siberia and northern Manchuria who grouped together to survive and thrive. Most of us live in association—as opposed to individually—as a way to manage and deal with the difficulties of life. This kind of diverse economic thinking seems to be rooted in actual projects through which common people seek to civilize the world's economy. Gibson-Graham's *A Postcapitalist Politics* shows that everywhere we turn, we find people organizing community-based economies, and that this is a reminder of the economic possibilities we can choose from. The use of mutual aid groups, particularly in the Global South and among rural communities and excluded minorities worldwide, confirms that diverse financial economies matter.

In *The Black Social Economy*, scholars and activists document how members of the African diaspora have had to self-exclude and physically isolate themselves because being in social contact with some white(ened) people has been unhealthy and harmful to their mental health.[8] Racial bias in employment that limits opportunities for qualified Black women is one example of deeply embedded inequities that affect people's well-being and their access to goods, reinforcing Kropotkin's finding that species gravitate toward mutual aid to depend on others for love, sympathy, and trust. But what happens when that love is not there? Black people living in majority-white or white-dominated societies have and will continue to make self-sacrifices: to first physically isolate and then to reach out to their community for mutual aid and support.

In *Collective Courage*, economist Jessica Gordon Nembhard outed—yes, made a truth known that was buried—that Black people's expertise in collectives and cooperatives during troubled times is remarkable.[9] It took her a decade to tell this story and correct the history of economic cooperation, which had previously ignored Black people's contributions. Gordon Nembhard explained that African Americans would turn inwards to ensure safety and to engage in cooperation to meet livelihood needs. For far too long, Black people have been overlooked as experts in cooperativism. In a recent study of W.E.B. Du Bois, economist Curtis Haynes Jr. also revises history to show how collectivity has been very much a part of the Black experience.[10]

My work on mutual aid focuses on rotating savings and credit associations (ROSCAs) and moves our understanding of coming together from one of necessity to pragmatic decision-making for our own well-being. Over the last ten years, I have spent hours talking with women known as "Banker Ladies"—African Canadian and Caribbean women who create community-driven financial cooperatives. They are collectively organizing as part of a centuries-long tradition of mutual aid groups working under constant crisis. These Banker Ladies have mastered how to pool resources to safely help each other, often with minimal contact with those harmful to them. Africans, West Indians, and members of the African diaspora know how to create community-based, inclusive cooperatives such as Susu, Partner, Sandooq, Box-hand, Hagbad, Equub, Osusu, and Ajo.[11]

I have witnessed hundreds of Black women in the Caribbean and Canada carry out such money cooperatives in difficult and desperate times.[12] In precarious economies coupled with racism and sexism, Black women—including Canadian women—are more likely to be under- and unemployed than men and white women. Many women in the African diaspora are used to surviving in an economy that excludes and discriminates against them, and it is for this reason that many choose to organize ROSCAs.

Black women engage in self-help and mutual aid groups because of what this sense of coming together brings them: a sense of duty to their social group, compassion, and love.[13] In their interviews with me, they have shared that they do this collective work, not only to meet their needs, but to press against exclusion and

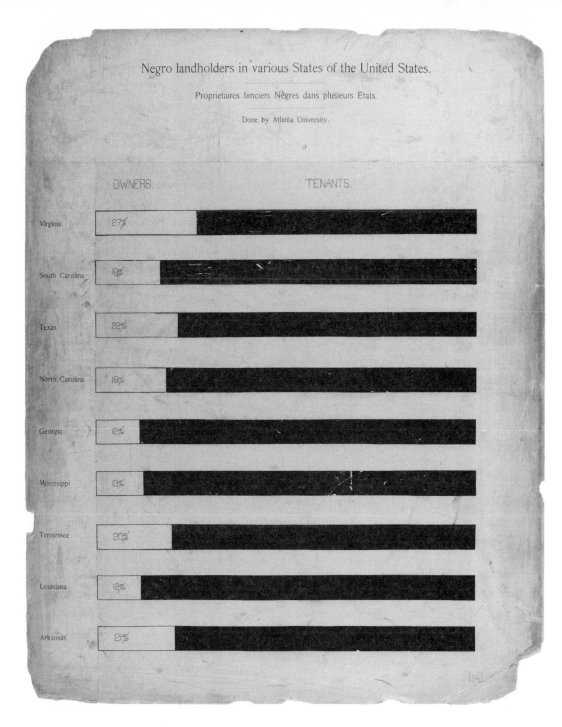

W.E.B. Du Bois *Negro landholders in various states of the United States*
From a series of statistical charts illustrating the condition of the descendants of former African slaves now in residence
in the United States of America, ca. 1900.

to beat the system by showing kindness, care, and love to alienated people. Banker Ladies create a space for people to connect and ensure that the funds are made available to members to cover funeral expenses, or to invest in inventory for their small businesses.[14] By pooling resources when they are excluded from bank loans, ROSCAs create an important lifeline in the Global South and among its diaspora in cities in the West.

In focus groups with hundreds of Banker Ladies in the Caribbean and Canada, women testified that they admire, respect, and trust these groups because it is the members who decide how to manage the pooled resources. This work is not directed top-down by so-called experts, but by everyday women who bring their locality together for discussions on how to build inclusive mutual aid systems. This practice of ROSCAs has been carrying on for centuries throughout the world, and it is a gift from the people of the Global South to the West.

Mutual aid is a way of life. It is a constant for the Global South and always has been, yet it is newly relevant to white folks. This is an opportune time to center current economic and public health crises on the work and experience of Black and racialized people. Black people know first-hand what it means to physically isolate and stay socially connected through mutual aid. We all have much to learn from the African diaspora, and especially the women, who remain steadfast in their commitment to mutual aid and to one another. We just need to follow their lead. ○

This essay was adapted by the author from "Mutual aid and physical distancing are not new for Black and racialized minorities in the Americas," which originally appeared in HistPhil *(History of Philanthropy, histphil.org) on March 24, 2020.*

Notes

1. Shirley Ardener and Sandra Burman, *Money-Go-Rounds: The Importance of Rotating Savings and Credit Associations for Women* (Oxford, UK: Berg, 1996). See also Caroline Shenaz Hossein, ed., *The Black Social Economy in the Americas: Exploring Diverse Community-Based Alternative Markets* (New York: Palgrave Macmillan, 2018).

2. J.K. Gibson-Graham, *A Postcapitalist Politics* (Minneapolis: University of Minnesota Press, 2006). See also J.K. Gibson-Graham, "Enabling Ethical Economies: Cooperativism and Class," *Critical Sociology* 29(2) (March 2003): 123-161 and J.K. Gibson-Graham, *The end of capitalism (as we knew it): A feminist critique of political economy* (Oxford: Blackwell Publishers, 1996).

3. Stuart Rutherford, *The Poor and Their Money* (New Delhi, India: Oxford Department of International Development/Oxford University Press, 2000).

4. Karl Polanyi, *The Great Transformation: The Political and Economic Origins of Our Time* (Boston, MA: Beacon Press, 1944).

5. J.K. Gibson-Graham and Kelly Dombroski, eds., *The Handbook of Diverse Economies* (Cheltenham, UK: Edward Elgar Press, 2000).

6. Caroline Shenaz Hossein, "Mutual aid and physical distancing are not new for Black and racialized minorities in the Americas," *HistPhil*, March 24, 2020.

7. Petr Kropotkin, *Mutual Aid: A Factor of Evolution* (Manchester, New Hampshire: Extending Horizons Books, 1976).

8. Caroline Shenaz Hossein, ed., *The Black Social Economy in the Americas: Exploring Diverse Community-Based Alternative Markets* (New York: Palgrave Macmillan, 2018).

9. Jessica Gordon Nembhard, *Collective Courage: A History of African American Cooperative Economic Thought and Practice* (PA: Penn State University Press, 2014).

10. Curtis Haynes Jr., "From Philanthropic Black Capitalism to Socialism: Cooperativism in Du Bois's Economic Thought," *Socialism and Democracy*, 32, Issue 3 (April 2019): 125-145.

11. Caroline Shenaz Hossein, "Rotating Savings and Credit Associations (ROSCAs): Mutual Aid Financing," in Gibson-Graham and Dombroski, 354-361.

12. Caroline Shenaz Hossein, "The Black Social Economy: Perseverance of Banker Ladies in the Slums," *Annals of Public and Cooperative Economics*, 84:4, 423-442.

13. Hossein, 2013.

14. See *The Banker Ladies* (2021), directed by Esery Mondesir and produced by Diverse Solidarities Economies Collective and Caroline Shenaz Hossein, at filmsforaction.org/watch/the-banker-ladies/

Lisa Trocchia

APPALACHIAN OHIO

Food Systems and COVID-19 through a Bioregional Lens

In what has become an extraordinary era framed by the coronavirus pandemic, communities everywhere are experiencing the inherent vulnerabilities of the dominant industrialized food system. The links between how and where we grow and process our food, how we can or cannot access it, how we consume food, what we waste, and the impacts of local-to-global food and agriculture policies are becoming increasingly visible. It's all connected. Recognizing that food systems function interdependently, it's important to understand the nature of these relationships. Food is not only what supports our own health and happiness; it is critically tied to the well-being of our communities, economy, national interests, and the planet as a whole. While examining relationships informs efforts to create a more just and sustainable food system, there are also examples of how rooting this work in a deep sense of place can enable food systems resilience in the face of crisis.

I was born and raised in Appalachian southeastern Ohio. Here, like everywhere, we face challenges from COVID-19 restrictions. Food insecurity and job loss are amplified by significant threats to the viability of farmers, food producers, food retailers, and restaurants. However, after over forty years of working cooperatively to create a strong local food system, responses to these issues are emerging in a networked, bioregional context, and in ways that speak to a history of living close to the earth through the spirit of cooperative culture. Within this foodshed, the promise of community-based food systems is rising.

What makes a bioregion? They are not composed of straight lines. They resist imposed borders and mapped boundaries; yet, they are created. Bioregions are established by relationships to land and water and between people, animals, and plants. They are formed through deep layers of interconnection, shared meaning, and experience. Within a bioregion, a "life-place," ecological and social diversity creates strength and resilience. Identity is balanced between cultivating this diversity and celebrating what is held in common. The concept of a bioregion insists on a dynamic interpretation and always reminds inhabitants, especially during times of crisis, to consider how ecologies are changing and relational.

My bioregion is tucked within the foothills that roll out from the western edge of the Appalachian Mountains. Located in the southeastern corner of the state, the roots of the area are buried deep in the coal seams and unglaciated clay soils of the Allegheny Plateau. The ancient Hocking River runs through the land, holding waters from many natural springs and

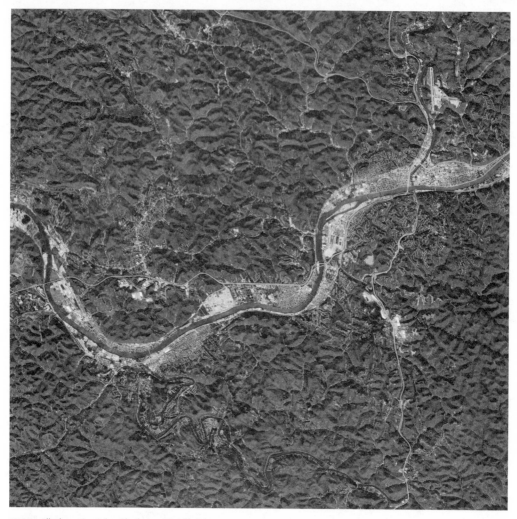

NASA *Allegheny Mountains, Charleston, West Virginia*

small tributaries as it meanders slowly to join with the mighty Ohio River where it divides the shorelines of Ohio and West Virginia. The visible legacy of the Indigenous Adena and Hopewell mound builder cultures can still be spotted rising from backyards and cultivated fields.

The topography differs greatly from the large cities and suburban sprawl that coexist with the flat, fertile fields of central, western, and northern Ohio. The contrast of abundant and diverse forests, sandstone rock outcroppings, poor soils, steep hills, and deep hollows found in the Appalachian Ohio bioregion has led to traditional differences in culture and opportunity, but also in the way food systems have evolved.

People have been engaged in a living relationship with the land in this place for thousands of years. The Indigenous people here were the first to cultivate wild foods, making it one of only seven regions in the world where local plants became the basis for a food economy.[1] Subsequently, European families were productive with diversified subsistence farming, but the hilly terrain eventually limited a wholesale transition to industrial agriculture. These limitations kept the ethos of small-scale family farming alive. After one-hundred-fifty years of extractive industry—salt, timber, clay, and coal—drained the region of its natural resources, exporting capital and impoverishing communities, those living here found a degree of food security tied to the retention of traditional foodways.

But it is more than the connection with home food production, or the skill of managing wild game, that began to characterize the region. Though proud of self-sufficiency, the tradition of mutual aid runs deep among the people living here. Expressions of cooperation speak to the memories and experience of having lived through hard times. It is this value that has emerged as a hallmark of the contemporary food system here.

Adding to this spirit, many "back-to-the-landers" made their way to the bioregion in the 1970s and 80s. Well into the late 1990s, as many as seventeen intentional communities were within a thirty-mile radius of each other. The influences of this population, interested in organic agriculture and cooperative structures, played a significant role in the eventual creation of the worker-owned cooperatives and nonprofit organizations that today are the backbone of the sustainable food systems work being done here.

Many communities across the country faced food shortages in the early days of the COVID-19 crisis stemming from the "just enough, just-in-time" industrial food system supply chain model. In my bioregion, however, nearly a half-century of food systems localization supporting Appalachian Ohio communities is demonstrating the advantages of a cooperative and community-based system. Prior to the pandemic, the *Utne Reader* referred to this area as one of three places in the world engaged in "some of the brightest ideas in the local food movement," and the region's food economy has been described as "a national model of sustainable community development." Ken Meter, one of the most experienced food systems analysts in the United States, characterized the cooperative and community-based strategies used here as "the best vehicle for rebuilding the American economy."[2] By fully activating these cooperative networks, COVID-19 challenges are actively being addressed within the food system by pivoting elements of existing regional markets to feed people through values-based value chains.

An example of these values in action relates to food access, always a high priority in the region. To address the additional challenges to food security presented by the coronavirus, many inspiring activities are emerging from existing and expanding networks as they become more diverse and focused on equity. Available funding for food access initiatives is being leveraged by nonprofit organizations working cooperatively to provide nutritious meals prepared by local chefs and restaurants using food purchased from area farmers and food producers. These are delivered by volunteers, and in some cases, by public school bus drivers, to rural residents in need. Restaurants restricted to pick-up or delivery options offer their customers the additional opportunity to purchase a low-cost meal prepared with locally sourced ingredients for distribution to others through the Neighbor Loaves and Meals program. The regional produce auction created an online platform for placing bids with cash-free, contact-free pick-ups and is working with non-traditional retailers, such as gas stations and convenience stores, to ensure there are locally grown fresh produce, meat, and eggs, available to those living in areas with limited access to traditional grocery stores.

These are just small examples of the many ways my Appalachian Ohio bioregion is responding systemically to food challenges: recognizing assets and honoring the legacies of biodiversity, self-sufficiency, and cooperation. It is in difficult times that values-based networks become visible. While the "survival- of-the-fittest" narrative is often valorized, when faced with immediate and complex challenges, it becomes clear that our most basic instinct as human beings is mutual aid. Survival has always depended upon cooperation. Pressed by the need to respond to a global health crisis, this may be the moment the kinds of values expressed in Appalachian Ohio will emerge within bioregional food systems everywhere. I hope as these place- and values-based food networks interlock, it will catalyze a wider shift toward national food systems sustainability, equity, and resiliency. ○

Notes

1. E.M. Abrams, A. Freter, *The Emergence of the Moundbuilders: The Archaeology Of Tribal Societies In Southeastern Ohio* (Ohio: Ohio University Press, 2005).

2. K. Meter, *Ohio's Food Systems – Farms At The Heart Of It All* (Minneapolis: Crossroads Resource Center, 2001). Commissioned by the University of Toledo Urban Affairs Center with funds from the Ohio Department of Agriculture.

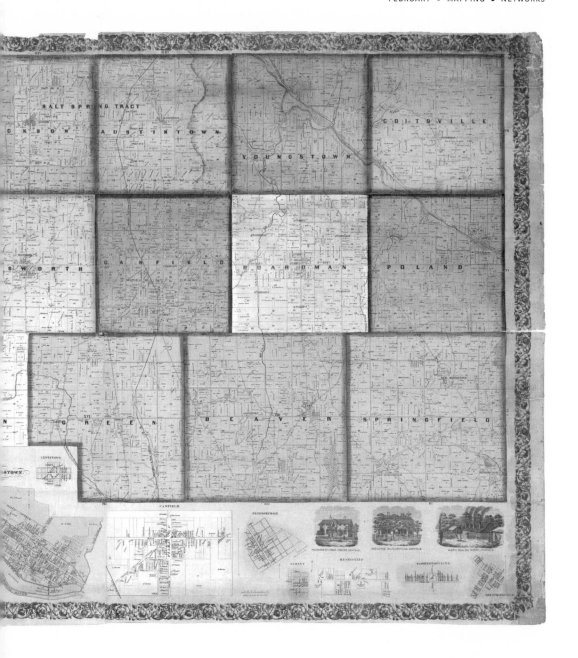

J. W. Canfield *Map of Mahoning County, Ohio, showing the original lots and farms*

77

Nina Pick

THE AFTERLIFE OF EARTH

After the fire comes rain. The prayers of grass
push up through the sidewalks.
Branches dance through abandoned houses
and rivers curl around the ankles of skyscrapers.
Walls are dismantled by the wind itself.
High above prison roofs padded with moss
monarchs ride currents in their wild migration.
The gray hawk flies across borders not caring
whose land it once was that she lands on.

Henry Weston

NEW ZEALAND'S ONE BILLION TREES PROGRAMME

In Aotearoa, New Zealand, we have a proverb: *Toitū te whenua, toitū te tangata.* If you look after the land, the land will look after you.

New Zealand's economy and environment are inextricably linked, which makes protecting our environment and supporting our transition to a sustainable, low emissions future one of the government's top priorities. It is focused on building a productive, sustainable, and inclusive economy that supports the well-being of all New Zealanders.

To do this, the government has a range of initiatives, including the One Billion Trees Programme. Launched in 2017 and managed by Te Uru Rākau, Forestry New Zealand, its goal is to double the current planting rate to plant one billion trees by 2028. The vision of the initiative is to deliver regional economic performance and jobs; support the aspirations of Māori, the indigenous people of New Zealand; help our country meet international climate change commitments; and support a range of other environmental outcomes.

The Approach

The One Billion Trees Fund and the Joint Venture Programme were established to provide funding to reduce the barriers to tree planting, build industry capability, and accelerate tree planting. This includes funding for landowners, organizations, and communities who want to plant trees or revert land to native forest; for groups or organizations with workforce development programs or projects that improve the way we plant or grow trees; and for joint ventures with private landowners.

The One Billion Trees Programme is focused on making it easier to plant the right tree, in the right place, for the right purpose. The approach is for trees to be integrated into the landscape to complement and diversify our existing land uses, rather than see large-scale land conversion to forestry.

The Benefits

While 80 percent of New Zealand was once covered in native forests, now forests cover just over a quarter of the country. Through One Billion Trees, we are seeing a resurgence in tree cover to protect and enhance our environment.

The programme is improving land productivity, income diversity, soil erosion and water quality, and native biodiversity. It is facilitating regional economic growth and opportunities for Māori to maximize the potential of their land and resources while providing access to work, training, and business opportunities.

Commercial plantation forestry, our nation's third largest primary export sector, also plays an important role in contributing to half of our one billion trees target. The opportunity through government investment is to enable the industry to diversify, innovate, and grow.

In the longer term, we will see wider changes to regulatory settings, including our Emissions Trading Scheme, which will further encourage tree planting across New Zealand.

Progress So Far

Between 2017, when the One Billion Trees Programme was announced, and December 2019, an estimated 150 million trees were planted by public, private, farming, and community groups across New Zealand.

By working with the farming and forestry sector, regional councils, and Māori communities, the One Billion Trees Fund has funded just over 16 million of these trees, fifty-one percent of them being indigenous species, and facilitated the creation of seventy-two full-time jobs.

There have been over two hundred and fifty direct grants to landholders, nearly fifty partnerships projects funded, and joint venture agreements to plant trees on more than five thousand hectares of land. More than twenty percent of funding has gone to Māori landowners or Māori-led projects.

Our current calculation of the potential for carbon sequestration of trees funded through the One Billion Trees Fund is approximately 1.5 million tons by 2030 and 6.9 million tons by 2050. This is the equivalent of the annual emissions of 3.5 million cars.

Project Examples

Direct Grants

Richard and Rebecca Riddell of Olrig Station near Hawkes Bay are incorporating trees into their farming mix to help protect and improve their soil and waterways while also helping New Zealand meet its international climate change commitments.[1] A grant has helped them plant pines in an area unsuitable for livestock and protect areas of native bush with stands of indigenous trees. They have also collaborated with their local council to plant poplars on erosion prone land.

Bob Webster and Carol Jensen purchased

their land, Waipuna Bush on the Banks Peninsula in Canterbury, with the aim of letting the land naturally regenerate. Funds from One Billion Trees are enabling them to restore their land with indigenous species, increasing biodiversity and enhancing natural landscapes. They are helping to build a natural bush corridor along the peninsula that will eventually bring back both birds and local plant species.

Partnership Projects

Hohepa Nursery is using funding to develop facilities where people with complex intellectual disabilities can be engaged in nursery production and participate as part of a wetland planting workforce. The project involves the production of thirty thousand eco-sourced native plants per year over the next three years. The area being planted is on the Ahuriri Estuary, one of the largest wetland systems on the east coast of the North Island, and an important ecosystem for many birds, plants, invertebrates, and fish.

Local young Māori men in eastern Bay of Plenty are benefiting from a partnership project that offers job opportunities and the chance to learn new skills in forestry. They can study towards forestry qualifications and are provided with support services to enable success. The young men are restoring seventy-three hectares of Māori-owned virgin native forest that was severely damaged by fire. The project is not only reducing erosion, improving water quality, and enhancing indigenous biodiversity, but also helping participants to re-establish a connection with their land. ○

Photos by **Ministry for Primary Industries**

Notes:

1. See "How trees fit into the farming mix at Olrig Station in Hawkes Bay," Ministry of Primary Industries, *YouTube*, November 27, 2019.
2. See "Tāne Mahuta 'Learn While You Earn' programme providing hope for rangatahi," Ministry of Primary Industries, *YouTube*, November 27, 2019.

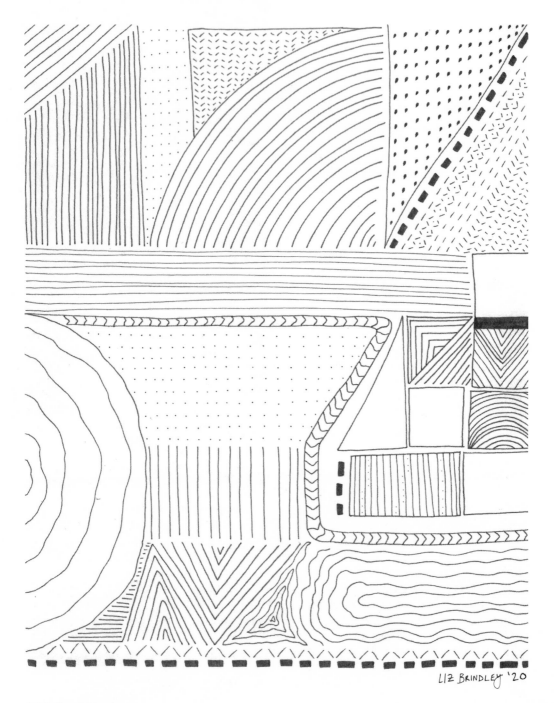

Liz Brindley *Patchwork*

Jo Ann Baumgartner

FOSTERING WILDLIFE-FRIENDLY FARMING AND RECOGNIZING BIODIVERSITY AS CRITICAL TO A FULLY FUNCTIONING FARM

We humans support what is beautiful and what we love. At Wild Farm Alliance that means songbirds singing out their names, fat bumble-bees busy sampling an array of gorgeous native flowers, and the majestic oaks towering over us—these animals and plants can and do live on farms.

There's a debate raging in the United States and especially in Europe, where eco-payments are made to farmers to protect rare species. It goes like this: Should farms be agricultural sacrificial zones, separate from protected wild places, or should we cultivate wildlife-friendly farms? In the sacrificial zones, a crop is grown fencerow to fencerow, every other plant is killed with herbicides or fire, and livestock are confined to small areas and are not allowed to graze. At Wild Farm Alliance, we maintain that from every perspective—beauty, functionality, and biodiversity—we must have both large protected areas and wildlife-friendly matrices on farms.

Agriculture comprises almost 60 percent of the contiguous US, and 40 percent of earth's landscape. As the human population grows and our planet heats up, it is imperative that farmers recognize the benefits biodiversity provides—particularly, pollination, pest control, clean water, and fertile soils. These benefits

can help farms be more resilient to changes in climate that are projected to cause increasing drought, heat waves, and floods. As farms support biodiversity, native species and ecosystems also become more resilient. Among the most important components of biodiversity, natural predators help control pests and diseases and help prevent over-browsing of plant communities. A biologically diverse farm, then, addresses the global extinction crisis.

A beautiful farm is one where the farmer makes a living not just on the land, but in collaboration with the diversity of the natural world. In so doing, that farmer has a richer experience interacting with nature (as opposed to solely growing corn and soybeans over and over). Customers who visit these farms have more memorable experiences, appreciating the beauty and eco-integrity embodied in the land.

Farmers with whom Wild Farm Alliance has worked on the central coast of California integrate nature at a high level. At Live Earth Farm, the songbirds sing throughout the day in breeding season—the insectivorous ones like the chestnut-backed chickadees eat codling moths so the worm that would have been in the apple instead helps to feed their young. The oak trees interspersed throughout the farm, especially those on steeper, untilled slopes,

support up to five thousand insect species. Hundreds of these are caterpillars which are critical high protein sources for nestlings. While honeybees (native to Eurasia) are in decline worldwide, bees native to North America have always helped with crop pollination. On this farm, conserved native habitat provides nectar, pollen, and nesting sites— pithy-stemmed plants like elderberry for tunnel-nesting bees, and open ground for ground-nesting bees. Corridors for all kinds of wildlife movement are protected on the farm. As their forty-some crops bloom, they give back to the pollinators and beneficial insects, supporting some of their food needs.

Of course, there are trade-offs to coexistence with nature. In Wild Farm Alliance's *Supporting Beneficial Birds and Managing Pest Birds,* we present over one hundred cases where birds provide pest control of insects, rodents, and pest birds in temperate agricultural crops. In a few crops, certain birds are beneficial most of the year as they eat insects, but can become pests for a short period during harvest. When this occurs, they should be supported in their beneficial phase, and discouraged while pests. Some birds, like American kestrels, help scare the fruit-eating birds from cherry and blueberry fields, though they mostly eat small rodents. Even with the trade-offs, the farm is viable and highly successful.

At nearby Deep Roots Ranch, they show that smaller farms can also be integrated with nature. They raise heritage breeds of cows, chickens, turkeys, and sheep. The secret to their success is that they are grass farmers— keeping the pasture healthy means matching the number of animals to what the grass can support. They have restored a muddy ditch to a clear running creek that runs through the middle of their property, supporting numerous songbirds and aquatic species.

To the east, at Phil Foster's Ranch, a restored riparian area stabilizes the river banks and provides a corridor that helps support wide-ranging predators which keep rodents in check. Huge native plant hedgerows support native bees and beneficial insects and birds. Transplanted crops are planted in minimally tilled fields to increase soil carbon storage, and the restored riparian corridor and hedgerows also store woody carbon. Trade-offs are managed with site-specific techniques. For example, the farmers plant the first few rows adjacent to one hedgerow with onion crops that aren't attractive to pest birds.

Much of US agriculture uses very different farming practices. Row crops are grown with "clean" edges and orchards are raised with cleared understories—space that could have supported pollinators and natural enemy insects. Instead, these mega-farms pay for pollination (beehives) and pest control (pesticides or beneficial insect releases). Animals are disconnected from their natural food sources and confined in small pens. If they are on pastures or rangeland, they are spread out on overgrazed land instead of moving through in tight groups, an evolutionary tactic for safety from predators. These conventional farms have a long way to go toward the better ecosystem functionality provided by biodiversity.

Farmers can diversify their operations through various wildlife-friendly practices. In Wild Farm Alliance's *How to Conserve Biodiversity on the Farm: Actions to Take on a Continuum from Simple to Complex,* we have identified a progression of activities that increasingly support biodiversity and the benefits it provides to the farm. Each farm has a unique set of circumstances and will begin at different plac-

es in the continuum, depending on its need and capacity for supporting nature. Whether the need is to build better soil health and clean water, ensure more complete pollination and effective pest control, or enhance habitat for wildlife, a farm can start with small steps or take big strides to integrate biodiversity.

Biodiversity Continuum

SIMPLE activities · COMPLEX activities
to maintain biodiversity · to improve biodiversity

1. Soil Life

Sustainable farmers start with building soil life to promote plant growth. Adding organic matter to the soil in the form of cover crops and compost supports a diversity of bacteria and fungi and builds soil carbon. Conserving untilled areas supports ground nesting bees, predatory beetles, reptiles and amphibians (herps), and birds, bats, and other mammals that consume crickets, beetles, and other ground-living insects.

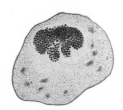

2. Soil Cover

In keeping soil covered, farmers do more than growing a healthy crop—they support above-ground complexity. Covering the soil with a crop, pasture plants, or non-invasive plants protects water quality, conserves soil, and captures carbon in plant tissues. Cover crops and plants also provide refuge for beneficial insects like damsel bugs and snakeflies, and for herps and ground-nesting birds.

3. Water, Nest, and Shelter Features
With these features, farmers spend a little more of their time supporting wildlife. By conserving waterways and creating puddles and ponds, they provide clean water for pollinators, natural enemy insects, birds, and mammals. These animals use the water for drinking, bathing, swimming, and nest-building. Farmers are also installing escape ramps in water troughs, building brush piles, and putting up nest boxes.

4. Flowering Plants
As farmers begin to care for non-crop vegetation, the amount of time and energy they expend increases, as does biodiversity's benefits. They support pollinators and natural enemy insects with a balanced and extended food supply, including a mix of plants that flower sequentially through the cropping season in field borders and pastures. These plants can provide cover for herps and food in the form of seeds and berries for birds and mammals.

5. Native Plants
Farmers have to seek out native plants because not every nursery carries them. They are worth the extra effort. In many ways, native plants are easier to care for than non-natives because they are adapted to local climatic conditions. They are further along in the continuum because they generally support more plant-eating insects and wildlife. Key natives like oak and willow trees are critical in nature's food web. Insect predators and parasitoids need alternate prey or hosts (plant-eating insects) when crop pests are not present. Herps, birds, and mammals rely on these insects, to a larger degree, depending on the species. Native plants are often used in hedgerows, windbreaks, and riparian areas.

6. Plant Structure and Composition
By increasing the numbers and kinds of trees, shrubs, forbs, and grasses along crop perimeters and interspersed through pastures, farmers provide food, cover, and nesting sites for many kinds of wildlife throughout the lower, middle,

and upper habitat canopy. The woody species store carbon. Farmers are also conserving grasslands, shrublands, woodlands, wetlands, and riparian areas, especially habitats of the highest conservation value that support rare species.

7. Corridors

With the addition of corridors, farmers link together habitat patches and pastures on the farm and connect them to wilder areas off the farm. Hedgerows, windbreaks, and riparian habitat can be linkages that bring pollinators and natural enemy insects and birds closer to the crops where they can be most effective. These corridors, especially when wide, provide safe passage for wildlife.

The most complex systems provide the greatest benefits for keeping wild species and ecosystems healthy, including those that are rare. The increased complexity also better buffers impacts from climate change. As more farmers adopt diversified farming practices, we will have a mosaic of food production areas that provide increasingly valuable ecosystem benefits.

Spending time in nature makes us feel good and is critical for our health. If farmers could share the sights, sounds, and smells of nature from their farms, it would undoubtedly widen their customer's enjoyment and support for their food. Music in the form of bird song triggers emotions that make us feel good. Smells do too. In forest bathing (*shinrin-yoku*), a Japanese therapy that reduces stress, forest sounds and smells are promoted. Why not "farm bathing," where a coppice of native trees is planted to provide shade for your lunching customers and workers, as at Swanton Berry Farm? Or a beautiful natural area on your farm conserved for family picnics, as Amish farmer David Kline has written about? Why not a path that customers take on their way to the farmstand, passing fields bordered with flowers and trees like those on Riverhill Farm?

Gastronomists believe we enjoy our food more when we identify with its origin—for example, people rate their enjoyment of eating shellfish higher when they are listening to sounds of the ocean. If a consumer liked eating a local, pesticide-free apple, they might love eating one that grew with native bees buzzing in its flowers, songbirds serenading and keeping it safe from pests, and creeks murmuring by. This is the next level of "buying local." It supports all kinds of "local" community members, farmers and wild residents alike, and incorporates the beautiful sights and sounds and necessary biodiversity of our world. ○

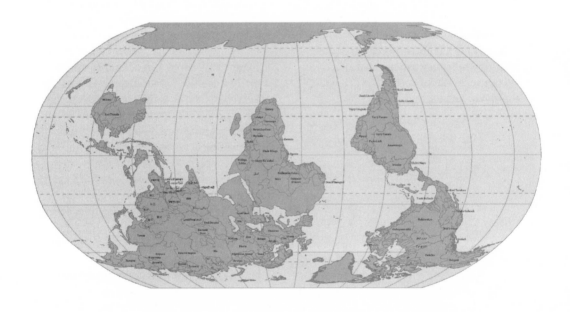

↑ **Jordan Engel** *Endonyms of the World's Watersheds*
It's an unfortunate limitation of mapmaking that there isn't enough space on the page to contain the vast diversity of geocultural knowledge that exists for each place. Here, our imperfect approach was to label each river basin or terminal lake with its name in one of the watershed's more widely spoken native languages.
→ **Carol M. Highsmith** *What looks like a series of puddles is all that's left, in many spots, of what looks like huge Goose Lake, stretching for miles from northeastern California into southern Oregon*

Watershed

Army Corps of Engineers

Staff members at Kua'āina Ulu Auamo

KUA'ĀINA
THE BACKBONE OF THE LAND

A Glossary for Shifting Paradigms in Land Use
and Biocultural Resource Management

*Na wai ho'i ka 'ole o ke akamai, he alanui i ma'a i ka hele e
o'u mau mākua?*

Who would not be wise, on paths well traveled by my forefathers?

—Reply of Kalaninuiliholiho Kamehameha II when praised
for his wisdom.

Kuaʻāina Ulu ʻAuamo means grassroots growing through shared responsibility. Below we present conceptual meditations on how we *hoʻo kahua* (build the foundation, principles, platform) as a community network-driven organization that uplifts transgenerational and place-based environmental wisdom and practices to care for Hawaiʻi's resources and people.

ʻĀina is land or earth—literally "that which feeds." *ʻĀina momona* or "fat lands" stretch across and uplift the ecological and cultural functioning of the entire watershed. This term is often used synonymously with visions of healthy, abundant, productive ecological systems that support community well-being.

Kuaʻāina are the grassroots, rural peoples of Hawaiʻi. These fishers, farmers, and caretakers are the backbone of the land. The stories and traditional ecological knowledge of these experts of land and sea are an inspiration to indigenous and local communities in Hawaiʻi and globally who advocate for self-determination and that which sustains us, our ʻāina.

ʻAuamo is a pole or stick used to carry burdens across the shoulders. In taking up the ʻauamo, kuaʻāina communities share the weight of carrying forward their vision for a better Hawaiʻi.

Ahupuaʻa is a watershed-scale land division, often extending from ridge to fringing reef. It is part of a bioregional management system that includes the larger *moku* (region, district) and *mokupuni* (island). Traditionally, activities within the ahupuaʻa included an exchange of resources between upland and coastal families. This place-based leadership mobilized the collective management of the commons from *mauka* (inland) to *makai* (ocean).

Loʻi kalo is an irrigated agricultural terrace for *kalo* (taro), a staple food and significant component of the Hawaiian diet, origin story, and our identity. The insightful and innovative practices of our ancestors is reflected in the design and construction of traditional agriculture and aquaculture systems that demonstrate an exceptional understanding of engineering, hydrology, ecology, and biology. These biocultural systems for food cultivation and natural resource management once supported close to a million people.

Loko i'a, or traditional Hawaiian fishponds, are an advanced, extensive form of aquaculture found nowhere else in the world. While techniques of herding or trapping fish can be found around the globe, this technologically unique system advanced the cultivation practices of *mahi i'a* (fish farmers) in Hawai'i. Traditionally, loko i'a served a sophisticated and essential role in protein production, producing an estimated three hundred pounds of fish per acre per year.

Kuleana means right, privilege, concern, responsibility. Contemporary usage implies doing your job, or certain title or rights in land, but the meaning is much deeper and speaks to the interdependence of rights and responsibilities and how we live in Hawai'i.

Konohiki means to invite ability and willingness. In the Hawaiian Kingdom era, the konohiki was often a *kaukau ali'i* (lower chief) who managed resources with the people of an area. In later times, the term became synonymous with land manager. Konohiki were measured by their ability to mobilize the collective endeavors of the community and facilitate 'āina momona. Its meaning, as set forth above, remains a concept and mindset for the future of Hawai'i.

Lawa means enough, sufficient, ample; to have enough, to be satisfied. This ethic of using just the right amount of resources for current needs ensures abundance for the future. In combination with the word *i'a* (fish), *lawai'a* (fisherman) have the experience to know what is enough and the foresight to maintain healthy, abundant fish populations. *Lawai'a pono*, to "fish righteously," means to fish Hawaiian, in ways that honor the values and traditions of our elders.

Mālama means to care for, attend, preserve, protect, beware, save, maintain. *Hui Mālama Loko I'a* network is a growing consortium of *kia'i loko* (fishpond guardians or caretakers) and stewardship organizations from fishponds across Hawai'i. This network emerged in 2004 as an opportunity for practitioners to empower each other and leverage their skills, knowledge, and resources related to fishpond restoration and management. There are many lessons to be learned from the practice of *mālama 'āina* (caring for the land/sea), especially in areas of climate mitigation, adaptation, resilience, resource management, and food sovereignty.

E Alu Pū means to "move forward together." It's a call to action—mimicking the movements of schooling *pualu* (a type of surgeonfish). E Alu Pū is a movement of community-based projects, families, and organizations involved in the stewardship of biocultural resources throughout Hawaiʻi. This intergenerational learning network gathers together annually to share lessons learned; support, teach, and mentor each other; and to nurture a productive space for growth and strengthened relationships. These kuaʻāina work side-by-side to reclaim physical spaces, improve resilience, and resurface indigenous knowledge, observation, management, and abundance in their places. They draw on practices developed over many centuries, innovate new ones, and are connected by grassroots values that drive their efforts toward a collective vision of ʻāina momona. ○

Note
Literal translations have primarily been sourced from *Hawaiian Dictionary* (1986) by Mary Kawena Pukui and Samuel H. Elbert, available online at wehewehe.org.

Kuaʻāina Ulu ʻAuamo (KUA) is a movement-
building organization that empowers grassroots communities across Hawaiʻi that
work to care for their environmental heritage
and achieve abundant, productive ecological systems that support community
well-being. KUA employs a community-
driven approach that currently supports
three networks of local, community-based
natural resource management initiatives
and indigenous practitioners: E Alu Pū,
Hui Mālama Loko Iʻa, and the Limu Hui.
Mahalo to contributing artists for this story:
Mark Lee (Holladay Photo), Scott Kanda,
and Kim Moa.

Tao Orion

WE ALL LIVE DOWNSTREAM

Private Timberland and the Politics of Enclosure

> We live in capitalism. Its power seems inescapable. So did the divine right of kings. Any human power can be resisted and changed by human beings. Resistance and change often begin in art, and very often in our art, the art of words.
>
> —Ursula K. Le Guin

It's late September, and I'm working out in my garden in the morning light before the dry heat of the afternoon settles in. Here in western Oregon, the transition time between late summer and early fall is precious—sweet, dew-kissed mornings unfurl into afternoons still warm enough for dips in the pond, or perhaps even one last trip to an icy river coursing down from the Cascade Mountains with veins of still-melting snow. Apples are ripening, pears are in the dehydrator, and peach slices fill jars in cool cellars. In the rhythm of the agrarian lifestyle, it's a high point of the year, a time of reaping the rewards of spring's scurrying and summer's tending, watering, and weeding.

On this day, another rhythm descends into my earnest attempt at bucolic bliss. The throbbing pulse of helicopter blades subsumes morning birdsong, quickens my own steady pulse. I watch the craft as it hovers over the clear-cut valley across from the small patch of land where I make my home. A fine whitish mist emerges from tanks secured to its flanks as it arcs back and forth across the slope. From one straight edge of the clear-cut to the other, back and forth, back and forth. Spray, turn, spray, turn, up and down the slope. I stand watch for about ten minutes, then head inside to check on my napping nine-week-old son. The pulse of the blades continues. From my kitchen window, I watch herbicide rain down on the barren land across the creek. I close the windows and watch the leaves on my newly planted apple trees—is the wind picking up, carrying the volatile herbicide onto the rich organic soil I've painstakingly tended?

Last winter, I woke at 4am to floodlights stationed on the top of the hill and heard the low rumbling of a feller-buncher working its way through the forest, followed by the short cartoonish whistle of a cable yarder as the choker setter indicated a log was set and ready for hoisting. I knew the spray was coming, that this helicopter would swoop into the valley, stealing its peace for an hour and changing its chemistry (and mine) for longer. It's just another step in a long line of processes culminating in plywood and 2×4s (as well as bare soil, dry creeks, decimated habitat, scotch broom, and massive carbon dioxide emissions).

In my time in this stretch of the upper Willamette watershed basin, I've borne witness to one ridge after another being cut, cleared, and sprayed. Oregon's notorious rain still most certainly falls on those slopes, but as it falls it no longer lingers upon infinite needles, runnels down rough bark, nor mingles with millennia of decaying duff before glissading through fractals of roots and finally, slowly, cleanly, making its way into the streambed to feed cutthroat trout, lamprey, heron, bear, eagle, and weasel. Calico Creek today runs dry one month earlier than when I arrived in this watershed fifteen years ago. Desertification happens on our watch.

As this devastation has unfolded, I've beavered up the channel by cutting small trees and branches and arranging them just so in the steep ravine, then watched as winter rains brought feet of sediment to rest behind these organic shapes. That precious soil is never going to move uphill again, so I'm happy to find it lodged behind limbs. In the cool pools bounded by these structures—whispers of floodplains past and future—tiny trout linger, attempting to inhabit the creek encoded in their memory as home and refuge.

No matter how much I attempt to rebuild and revitalize this minor artery of the Columbia River Basin's vast heart, what happens upstream—cut, spray, and cut again—is outside of my jurisdiction as a citizen. I don't get to vote on the forest management practices that affect my life and livelihood, and neither do the trout or weasel. Federal laws protect the rights of timberland owners to further their economic interests at the expense of the local commu-nity, the human and non-human species who share this sun-kissed valley. Such laws have a long history in the warp and weft of the colonial-capitalist nation-state struggling towards democracy now known as the United States.

The structural framework used to strip citizens of their ability and inclination to participate in land management is intertwined with the origins of capitalism, the enclosure of the commons, and the resultant dislocation of people from place and means of livelihood. If we are to envision different ways of relating to each other, the lands we inhabit, and the lives that we're intertwined with, it's necessary to hold a magnifying glass to those threads as we work to unravel them.

While social and economic inequalities existed prior to and outside of the development of capitalism, this economic system's rise provides important context about the situation in my little valley, and those in other valleys, ridges, plains, and mountains all over the world. The capitalist enterprise first gained traction in western Europe at the end of the 14th century. In *Caliban and the Witch*, historian Silvia Federici traces the social and economic upheavals caused by the enclosure of common lands throughout the Middle Ages in Europe as a central driver of this transition.[1]

There, most land was held in manors and estates by relatively few elite people. The vast majority of the population lived on these lands as peasants and serfs, often in exchange for agricultural goods or crafts. The relationship was far from equal. Lords of the manor often dictated marriage arrangements and other personal affairs of the people inhabiting the land they ostensibly owned and managed. Those people, however, managed their own home gardens, and large portions of land were often managed as a commons from which any person could

Albrecht Dürer *The Expulsion from Eden*
This scene of the expulsion of Adam and Eve from the Garden of Eden evokes the expulsion of the peasantry from its common lands, which was starting to occur across Western Europe at the very time when Dürer was producing this work.

sports, games, feasts, and gatherings were, according to the ruling class, proof of laziness and poor work ethic.

The subsequent process of enclosure—privatization and prohibition from public use and stewardship of the commons—caused uprisings and revolts throughout western Europe until the 19th century. Access to land for livelihood became increasingly scarce throughout the 15th century and accelerated through the 19th century. Those without a land base flocked to burgeoning cities in search of wage-based work. Mills that processed wool from sheep pastured on what had been common land became one of the first scaffolds of the British Empire, whose rise to prominence as an economic power drove the process of colonization throughout the world. Enclosure and dispossession from the land base severed familial, social, and economic connections. Church and state consolidated to address these emerging challenges; police, jails, workhouses, and orphanages sprung up throughout Western Europe. Witch hunts swept through Western Europe at this time, wherein hundreds of thousands of people—mostly women—were executed for behaviors that did not conform with the social order prescribed by a ruling class interested in concentrating land, money, power, and control over the emerging wage-based workforce. As this emerging social order took shape, newly created wealth financed colonial conquests in the Americas, Africa, Oceania, and beyond.

Wherever they went, European colonizers brought their worldview with them. It was informed by the brutal social and economic reality encoded over the past centuries. Their concept of private property and accumulation of wealth was engendered by capitalism's unceasing appetite for human and natural

access important livelihood needs, including firewood, rivers for fishing, land for hunting and pasturing livestock, and building and craft materials. Commons-based management bolstered local and regional decision making, celebration, and solidarity, as people came to intimately know how to live together on land and form community around its care. As this peasant class grew in population and power, many within it began challenging the authority of the manorial class. Elite landowners lamented what they perceived as a lack of discipline among the peasantry, who appeared to work as little as possible toward the tithes they owed for living on the land. Their seasonal festivals,

"capital." Historian Howard Zinn recounts:

... the frenzy in the early capitalist states of Europe for gold, for slaves, for products of the soil, to pay the bondholders and stockholders of the expeditions, to finance the monarchical bureaucracies rising in Western Europe, to spur the growth of the new money economy rising out of feudalism, to participate in what Karl Marx would later call "the primitive accumulation of capital." These were the violent beginnings of an intricate system of technology, business, politics, and culture that would dominate the world for the next five centuries.[2]

Colonizing interests in Africa, Australia, New Zealand, Polynesia, and the Americas forcefully seized lands that had been continuously inhabited and tended for millennia with diverse strategies of tenure and use rights, including commons-based management. As British interests set their sights on the vast resource wealth present in the rivers, forests, prairies, and wetlands of what would become the United States, it was the notion of private property that laid the framework for dispossessing Indigenous people from their land. John Winthrop, leader of the 1630 migration of English Puritans to what would become known as the Massachusetts Bay Colony, reasoned that Indigenous people had no legal rights to the land because their claim to it was based on a different type of land tenure than that practiced in Europe.

That which is common to all is proper to none. This savage people ruleth over many lands without title or property; for they enclose no ground, neither have they cattle to maintain it... And why may not Christians have liberty to go and dwell amongst them in their wastelands and woods...? For God hath given to the sons of men a twofold right to the earth; there is a natural right and a civil right. The first right

was natural when men held the earth in common, every man sowing and feeding where he pleased: Then, as men and cattle increased, they appropriated some parcels of ground by enclosing and peculiar manurance, and this in time got them a civil right.[3]

Winthrop's argument, which became the basis of dispossession of land throughout the nascent US and beyond, was that since Indigenous people held land in common, they had only a "natural" right to it, but not a "civil"—or legal—right. There were no obvious boundaries to people's use of the land, he said, therefore it wasn't being used in the "correct" way. But prior to English contact, Dutch, French, and other European explorers, travellers, and colonists had published maps and accounts noting established villages where distinct Indigenous tribes cultivated corn and other crops on a large scale; maintained hunting grounds, orchards, pastures; and constructed elaborate fishery catchments and weirs, among other developments in rivers and lagoons.

It was abundantly clear that the land was tended and productive, though not fenced and "owned" as it was in Europe (where the concept of "private property" was normalized only after centuries of dispossession, genocide, and uprising). Indigenous people had no titles to land and water resources printed on paper proclaiming personal ownership; their understanding was based on a different set of rules and values.

Four hundred years and countless enclosures later, Nobel Prize–winning economist Elinor Ostrom researched commons-based resource management systems in use throughout the world. She found that while they are not without their difficulties, each represents a unique set of ways in which people work together to ensure the longevity and health of the resource base that they all rely on for their livelihood needs.

What Ostrom elucidated, and what practitioners of commons-based approaches to land management understand, is that what makes the commons work isn't the land itself, rather the process by which people with a stake in its vitality engage with it. As commons researchers David Bollier and Silke Helfrich note:

Commons are not things, resources or goods; they are an organic fabric of social structures and processes. They may be focused on managing a certain resource—land, water, fisheries, information, or urban spaces—and those resources may have a strong influence on how governance structures and economic production occur. But excessive attention to the physical resources or knowledge that a commons relies upon can distract us from its beating heart: the consciousness of thinking, learning, and acting as a commoner.[4]

Commons aren't a free-for-all; historic and contemporary commons are notable for their definition of who, how, when, and why interaction with this land is deemed appropriate by the group of people most affected by its use. Ostrom's extensive research on the commons found eight general principles shared by practitioners throughout the world:

1. Define clear group boundaries.
2. Match rules governing use of common goods to local needs and conditions.
3. Ensure that those affected by the rules can participate in modifying the rules.
4. Make sure the rule-making rights of community members are respected by outside authorities.
5. Develop a system, carried out by community members, for monitoring members' behavior.
6. Use graduated sanctions for rule violators.
7. Provide accessible, low-cost means for dis-

Nina Elder *Rot Pile*

Indians farming on Fort Peck Reservation, Montana, circa 1918. Fort Peck is home to numerous bands of the Assiniboine and Sioux nations who once lived from and with the buffalo. The Lakota Sioux view all life forms as related—human and nonhuman, plant and animal, water and rock. This includes the sacred buffalo, slaughtered to near extinction by white capitalists working for the fur trade and the westward expansion of the US.

pute resolution.

8. Build responsibility for governing the common resource in nested tiers from the lowest level up to the entire interconnected system.[5]

Thus the commons exist as structures of relationships among people who are committed to working together to manage shared spaces for meeting their own needs. They are by nature locally based, constantly adapting to changes inevitable in organic social-ecological systems.

Unfortunately, much of the discussion about commons-based land management comes from Garrett Hardin's 1968 infamous essay, "The Tragedy of the Commons," which describes hypothetical pasturelands driven to the point of collapse by the continued additions of cattle by people who manage the land in common. Hardin argues that individual self-interest leads to the eventual decline of commonly managed resources because each person who uses the resource seeks to maximize their own benefit. In Hardin's example, graziers are always interested in adding one more cow to the pasture even if it means poorer quality pasture in the long term.

Operating under the assumption that individuals act solely within the context of their own self-interest, the "tragedy of the commons" unfolds. The concept is based on the idea that people are free actors out to maximize everything they can from a given piece of land for their own personal gain. It negates the existence of social structures or norms of use enforced on a local or regional level by the people who make shared use of the resource. Hardin never did any research on historic or existing commons-based management systems, unlike Ostrom, who note that people all over the world tend towards finding ways to work together to manage the land and resources they love and need to survive.

The "tragedy of the commons" perpetuates the myth that private property is the most efficient and logical means by which people relate to the world that sustains them, that "market forces" will dictate right and good action that benefits all. It has led to legal frameworks in states like Oregon, whose Right to Farm Law enables individual landowners to make the decisions based on their own interests. Ironically enough, the very outcome that Hardin sought to warn against exists in the framework of a reality informed by the concept of private property that we are embedded in today. Each property owner is bound by the confines of their own self-interest and often driven by economic aims that repeatedly fail to consider the greater community. Just as Hardin's imagined tragedy fails to grasp the dynamism and potential of the commons, the world as imagined by capitalism fails to account for the downstream effects—the externalities—that so profoundly shape all of our lives. The process of enclosure continues, and our fundamental commons—the air, water, soil, climate, human, and non-human bodies that we all share and rely on—are treated as reservoirs for residues, effluent, debris, "waste," leaks, spills, drift. Our bodies and ecosystems house the "waste" of capitalism, which we are forced to deal with in isolation and without recourse or recompense. We all live downstream from these ongoing enclosures.

Understanding the commons as our collective heritage should inform how we engage in the care-full process of reckoning with the violence and inhumanity at the center of the economic system that we live in today. We must uproot its foundational ethics of racism, sexism, competition, accumulation, and dispossession by charting a course towards a future where commoning in our forests, fields, cities, wetlands, and prairies is the foundational means through which we relate to each other and the world. The edges of this global economic system, with its unprecedented power, breadth, and scope, are fraying. Commoning kindles potential for weaving a new story from the ashes of capitalism's necessary demise.

So, still, the forestland upstream and upwind of where I reside is bound in a web of rules and regulations that aim to preserve the sanctity and longevity of private property as a basis for the colonial-capitalist enterprise. But I envision a world in which citizens have the opportunity and responsibility to engage in management in and on the land that makes their livelihoods possible. People should have the natural and civil right to participate in the land that sustains them, in the watershed that nourishes them. To speak and act for the benefit of beaver, heron, weasel, and bear. To know a life well-lived by ensuring crystal clear creeks flow through the summer, where cutthroat trout build redds, and smolts learn the scent and shape of the water that will always welcome them home. Strong-grained wood will warm hearths and build community homes nestled in valleys whose fertility and diversity is protected with carefully stewarded, canopied forests. Herbicides and helicopters, the unnerving pulse and the mystery mist, are relegated to the compost bin of bad ideas and worse outcomes. In their place, possibilities. ○

Notes

1. Silvia Federici, *Caliban and the Witch: Woman, the Body, and Primitive Accumulation* (Brooklyn: Autonomedia, 2004).

2. Howard Zinn, *A People's History of the United States: 1492-Present* (London: Routledge, 2015).

3. John Winthrop, "General Considerations for the Plantations in New England, with an Answer to Several Objections," *Winthrop Papers, vol. II* (Boston: Massachusetts Historical Society, 1931), 120.

4. David Bollier and Silke Helfrich, "Overture," in David Bollier and Silke Helfrich, eds., *Patterns of Commoning* (Commons Strategy Group and Off the Common Press, 2015).

5. Elinor Ostrom, *Governing the Commons: The Evolution of Institutions for Collective Action* (Cambridge: Cambridge University Press, 1990).

Brett Ciccotelli

RECONNECTING RIVERS, LINKING GENERATIONS

Protecting and Restoring Maine's Wild Fisheries

Just before the new year, in the kitchen of an old farmhouse way down off Route 1, an old-timer's eyes light up at the mention of a long forgotten winter fare: the tommy cod, or frostfish. That glint is common of folks of a certain age raised among the brooks, coves, and rivers around here. Whether pulled frozen from cold storage in a snowbank (tomcod), freshly dipped on a cool spring night (smelts), hooked in early summer (salmon or brook trout), or brought down from the attic (smoked alewives), much of what showed up on the table in coastal Maine first came up the watery threads of connected waterways.

Talk of cold nights and the fishy past leads out to the barn, where an eel spear, with its bent back tines for pulling hibernating eels from the mud, comes down from a hook on the wall for the first time in half a century. With the spear in hand, stories and a map of home defined by food and fishing slowly run out. An old cedar smelt ice fishing hut—now used to brace a few cords of well-seasoned firewood—brings to mind east winds that broke ice flows off shore and sent fishermen floating across Union River Bay. Old wooden boxes yield receipts and shipping labels from faraway fish markets. These yellowing pieces of paper show the path of fresh fish from Downeast Maine by railcar or

stagecoach to hungry families in the city. The tools and stories tell of hard, proud work.

What changed? I'm told people don't live off the land any more or that it's easier to buy food up at the grocery store.

This is only part of the story. In many places the fishes are gone. Dams, overfishing, pollution, and climate change have broken the link between rivers and the sea. The generations that knew the best place (the brook up behind the old homestead) and best time (not before Uncle John's birthday) to gaff a tommy cod for chowder on Christmas night are passing on. Without the fish, they've had no reason to hand down the knowledge that was handed down to them.

These migratory fish fed more than the people along the banks. Their return each year supported communities and enriched the land and water. Though some of Maine's migratory fish species can spawn more than once in their lives, returning to the ocean to repeat their migration in future years, many die or are eaten during their runs each year. Eagles, otters, snapping turtles, and wolves thrived off the bodies of the fish and their eggs, carried inland by the millions each year from the fertile waters of the Gulf of Maine. The bodies of those fish that do not survive to spawn again decay in the

CODFISH.

water or are carried off into the forest, the bony remains fertilizing the water and landscape with essential minerals and nutrients gathered in the sea. Sea lamprey dig spawning beds that clear the river bottom of sediment and prepare places for salmon to leave their eggs. The loss of these runs rob the land, water, and communities that depended on their predictable returns.

Just as the destruction of river fishes hurts the land, the Gulf of Maine also starves for want of fish. Codfish and halibut once waited at the mouths of rivers for the outmigration of the billions of young fish born upstream. The rivers supported unimaginable nearshore fish populations that sustained the Wabanaki people for millennia and tempted Basque fisherman across the wild North Atlantic.

We know what works to bring many of these species back home and reset the seasonal migration clock. When we repair broken connections between up and downstream waterways, fish begin to move back to where they belong. It always works. We see millions of river herring, shad, and sea lamprey return to the Penobscot, East Machias, St. Croix, and Kennebec Rivers following dam removals and the reopening of historic passageways. We watch Atlantic salmon begin the long road to recovery on the East Machias River. There, an innovative conservation hatchery and restoration program run by the Downeast Salmon Federation treats fish like fish and rivers like rivers. This model can replace the traditional factory and laboratory hatcheries that, combined with the destruction of river habitat and dam building, have pushed Atlantic salmon in the United States to the brink of extinction.

TOMCOD.

Eel

Thankfully, we live in a world where there are still rivers and brooks that run with the abundance that put that glint in the old-timers' eyes. There are also rivers being given new life—with some restored places seeing more fish than any time in the last two hundred years. These fishy corners of the world inspire families to head out on dark nights to look for wild food and a link to the past. Healthy runs give a new generation reason to spruce up the old smelt fishing hut in the barn and sharpen long-unused gaff hooks for eeling or frostfishing.

These places, where the fish still run and the passion to look for them exists, can inspire the spread of healthy fisheries back to all the streams where they belong. The rivers of Maine and Eastern North America once provided homes to shad, salmon, alewives, blueback herring, tomcod, eels, sea lamprey, sturgeon, striped bass, rainbow smelt, and brook trout. Rivers around the world were darkened each year with a diversity of fins difficult for us to imagine. The rivers' abundant past can be the future.

When we do our part, the fish come back. There's hard, proud work involved. By sharing in that work, the sight of those first fish each year will mean that much more. So wherever you are, go out, find your river, and see if you can help the fishes. ○

Rob Jackson

SPRING PRUNING

I shape a plum tree first,
leggy in its second-year growth,
clipping the central leader a foot or two
above the first ring of branches,
in turn trimming each of those
a shoe's length from the main stem
—just past a bud and at angle instead of flat
to ensure water drains.

Where disease struck the apple and peach trees,
I cut more, severing branches stricken
with fungal rust, fire blight, or virus,
disinfecting the pruning shears with alcohol between cuts.
I trim extra to be safe, mindful
of things newly pruned by virus from our lives:
unnecessary flights, shopping from boredom,
nights playing music with my son,
our elders and begetters.

Quinn Riesenman

MID-MARCH, 2020

I go to the grocery store to buy beer and I see an older woman exasperated by the orange juice void. The fluorescent bulbs behind the bottles, which I have never before noticed, cast a pale shadow of her legs and cart onto the checkered linoleum. She looks at me with a tired face, like, I can't believe this.

Back in my car, I sanitize my hands with alcohol wipes that have lived in my glove box since 2007. My beer is on the floor. I suppose I will sanitize it when I get home. I wonder if that's enough time for the virus to crawl or writhe or scuttle or move however it does up the passenger seat, over the cup holders, on to my leg, up my torso, along my face, into my nostrils, and down into my lungs.

In the parking lot, the orange juice lady opens up a truck door. I can see her husband behind the wheel. He cranes his head back to see what she got and what she did not get. As she loads the groceries into the backseat I can hear her listing with consternation all the products that had presumably been available every day of her life until two weeks ago. Orange juice, cheese. Toilet paper, of course.

It's sunny in the parking lot, but as I start my drive I find myself inside a cloud of snow.

I drive back to the house I live in, on a piece of land owned by the state of Colorado and managed by Fort Lewis College, called The Old Fort. Although I am originally from Colorado, I moved here in early February from California, where I'd apprenticed on a farm on the central coast for a year, to take part in a farmer-in-training program at The Old Fort. I got hired as

a cashier at a bakery in Durango, which I hoped would structure the few months before the start of the program in early May, but with the recognition of the virus's severity spreading, the bakery has postponed my scheduled training until things "clear up." Based on predictions in the news, I am beginning to wonder if I will ever have the experience of selling a baguette.

Aside from a few hours of weekly work trade at The Old Fort, I have no other obligations. With the statewide shelter-in-place order, even my attempts to feel and appear busy by going into town to write in the lamplit corners of coffee shops are impossible.

Suddenly cast into unasked for and unpaid free time, friends I speak to on the phone express worry and anxiety. Worry for money, for their safety, and for the safety of their loved ones. Worry, just for the sheer unbelievable feeling of it all, with death toll predictions for the United States alone in the millions. And anxiety, trying to figure out how to fill amorphous days at home.

I certainly feel waves of these worries, but I mostly feel relieved. I figure I can live well off my savings for long enough to stave off economic pressure until the farming season starts. Living alone in a house on a farm in a rural area with access to fresh greens and miles of open country to wander around in, I feel confident I can avoid both the virus and the boredom of quarantine. I embrace the socially sanctioned responsibility to stay at home; no longer do I need to feel guilty about avoiding social interactions, it is now My Duty.

In other words, by actively and consciously cashing in on my privilege, I have placed my-self in an environment in which feeling relief is possible.

Years of consuming post-apocalyptic video games, movies, and books combined with a basic understanding of the matrix of global crises currently unfolding on the planet has

contributed to this relief. Between our fic-tions and realities, deep tensions breed on a planetary scale. I know you feel it too. Many offer, and some embody, what they see as solutions—I became interested in farming as one such solution—but this tension seems only to be intensifying. At this early stage, I have hope this pandemic will act as a catalyst for larger-scale processing that helps us transition

Alli Maloney *Yuhaaviatam/Maarenga'yam (Serrano) Land*, 2018

of snow cap large rocks. Freshly broken cottonwood limbs dip at weird angles into the low flow. I step awkwardly over layers of multicolored stones. Ice cracks in the silence between the distant exhalations of semi-trucks hauling coal up and down the highway.

This slow, constant trickle of water brought this all here, all these stones. In the spring, this trickle roars, tilting trunks and boulders, and bears a single grain of sand from each stone with it. And somewhere downstream: beaches of sand blinding in the summer brilliance.

Have you ever found a stone on the beach or in the mountains or beside a river—a stone whose particular features remain after millennia of movement, erosion, water flow, hinting its distant origins—and thought you might be able to trace the entire history of the earth, if you had enough time to read it? Did you slip it into your pocket, or did you set it back down thinking, there's still more yet?

I worry about labels like "spiritual bypassing," "escapism." But here, take a look. There's no escaping the unspoken side effects of prolonged anxiety, nor the worry coursing its way through your veins. Down there in a clear pool on the edge of the river, settled reddish layers of last year's cottonwood leaves dancing, cold as it is. ○

into the more dynamic, enlivening systems of response that we consciously or unconsciously crave. Hence, a kind of relief.

But here in what everyone's calling the eye of the storm, what do I do with this complicated flavor of relief? How do I even know if this relief is appropriate or realistic?

It's cold and still when I leave my house with my walking stick and cross the highway. I pass a fenced-in electrical station and hop over the barbed wire. No snow falls. I walk east toward the river. I walk along La Plata. Islands

THE SARDINE INDUSTRY.
Shore herring weir near Eastport, Me.; the common form of brush weir. (Sect. v, vol. i, p. 591.)
From a photograph by T. W. Smillie.

The weir is a technology common to many cultures who have observed the circling swirl of schooling fish. Fish swim into the circular enclosure—built from sticks, branches, or fish nets—and then cannot leave. The benefit of a weir system, apart from not needing to go out trawling in large boats, is that fish can be released if they arrive in excess of need, allowing fishermen to be in more intimate relationship with the fish and their level of abundance.

—Severine von Tscharner Fleming

Lia McLaughlin, USFWS *Potter Hill Dam Fishway*
Hurricane Sandy resilience projects will assess the Potter Hill
Dam Fishway on the Pawcatuck River and remove the Shady
Lea Mill Dam on the Mattatuxet River in Connecticut and
Rhode Island.

Emily Vogler

BUILDING
WATERSHED
DEMOCRACIES

Five years ago, I was asked to join a team of researchers on a National Science Foundation grant looking at aging dam infrastructure in New England.[1] Coming from the West, hearing "dam" made me think of large, monumental, federally owned, hydropower dams such as Glen Canyon Dam, Hoover Dam, or Grand Coulee Dam. But as I started to visit dams throughout New England, I realized that we were dealing with a very different type of dam.

There are over fourteen thousand dams in New England. The majority are small "legacy dams," only five to twenty feet tall, built hundreds of years ago to power early colonial grist mills and later industrial textile mills. Many no longer serve their original purposes and are coming to the end of their life cycles. In its 2017 Infrastructure Report Card, the American Society of Civil Engineers gave the nation's dams a 'D' grade, indicating the lack of maintenance and poor condition of many of the dams.[2] Aging infrastructure, shifting climate regimes, and large storm events have heightened these concerns, as an increasing number of dams are at risk of breeching and threatening downstream communities. A case in point: as I began this essay, the Edenville Dam in Midland, Minnesota, was overtopping and eleven thousand downstream residents

were being evacuated from their homes.[3] In addition to addressing these safety concerns, future decisions about these dams present one of the greatest opportunities for large-scale restoration of New England rivers since the Clean Water Act became law almost fifty years ago. These decisions have the potential to improve habitat connectivity for endangered and threatened migratory fish such as salmon, herring, shad, and eels, and to improve water quality and restore the flow of sediments and nutrients that support critical freshwater and coastal habitats.

Since the 1990s, there has been a push by environmental organizations and public agencies to remove old dams to restore the ecological and hydrological connectivity of rivers.[4] From an ecological, economic, or safety perspective, it makes sense to remove legacy dams. But many policymakers and agencies have been ill-prepared for the unexpected resistance from community members protesting their removal.[5] Dams, and in some cases their associated reservoirs, have become a significant part of individual and collective sense of place in rural New England communities. Dams and reservoirs are landmarks within an otherwise unstructured forested landscape; they are places where people grew up fishing with their grandparents, landscapes people drive or walk by every day, and features that have led to higher property values. In some cases, the dams are on the town seal in recognition of the village's colonial and industrial history. To many, these dams are seen as a symbol of cultural and regional identity.

Public meetings to discuss the future of these dams are often controversial, with people coming out to defend either side and argue to either keep or remove the dam.[6] At one such meeting, a gentleman from the Swift River watershed in Massachusetts said, "If you kill the dam, you are killing a part of me."[7] These complex social dimensions have resulted in the delay or cancellation of over fifty dam removal projects in New England.[8] Beyond the threat to cultural landscapes and sense of place, communities are also resistant to outside agencies and authorities, who they feel come into their community, often with the funds and resources to remove the dam, and ignore local concerns, values, and desires.[9]

These controversies raise important questions about how the public should be involved in decision making about the management of rivers, water, and other common pool resources. Many people, myself included, advocate for environmental democracies, where communities have the right to make decisions about the future of their surrounding environment and water resources.[10] However, there are limited examples of how to actually achieve this in our complex contemporary society. If our goal is community control, we have to define "community" and include all voices in the decision-making process. How do we ensure that Indigenous communities, such as the Penobscot Tribe in Maine or the Wampanoag in Southern New England, who traditionally depended on the rivers as fishing grounds, are

heard within the predominantly white cultural context of rural New England? How are the needs of non-human species such as salmon and herring factored into decision making? How do you build the participation necessary to be representative of the broader community and not just capture the opinions of outspoken individuals? And lastly, are environmentalists prepared to accept the outcome of a community process, even if it does not lead to "better" environmental outcomes, like dam removal?

In response to these complex questions about how the public is brought into (or excluded from) the dam decision-making process, I have worked with students and colleagues from the Department of Landscape Architecture at the Rhode Island School of Design, where I am an associate professor, to design a process for engaging communities in a discussion about the future of their dams. Integrating methods from community design workshops with an environmental decision-making method called Structured Decision Making, we aimed to create a structured way for communities to discuss their values as well as scientific facts and evaluate the trade-offs around various future scenarios. Because many New England dams are small, there exist a range of approaches that can significantly improve habitat connectivity while allowing the dam and/or impoundment to remain in place. One option, the nature-like fishway, involves raising the entire downstream width of the riverbed to gradually meet the height of the dam. Another option is to construct a new river channel to "bypass" the existing dam. Alternatives like these help communities move beyond the binary choice of keeping or removing the dam and can achieve multiple objectives, from improving fish passage and water quality to maintaining recreation on the reservoirs to preserving historic dam structures. In addition to these alternatives to dam removal, we are exploring options for how to use design to maintain or strengthen sense of place when dams are removed, such as installations that mark sites of historic dam structures, managing vegetation to preserve open views, or maintaining the aesthetic experience of a dam through constructed water features.

One of the results of our work is a decision-making toolkit that includes a set of "trade-off" cards to help individuals discuss their values, a set of "alternative" cards that describe and help visualize the alternatives, and a matrix that compares the social, ecological, and economic impact of the dam alternatives. The toolkit relies heavily on visual graphics to help make

scientific reports, quantitative data, and future alternatives more succinct and accessible. It is currently being tested in various communities in New England and will soon be available to download online.[11] In one post-workshop survey, a participant commented, "I was surprised how willing I was to change my mind." Another acknowledged the need for compromise, stating "the historic value is something I would be willing to compromise on."

As I started writing this essay, the Edenville Dam was breaching, and as I ended, streets across the United States were flooded with protests against racist violence and police brutality. The current problems with aging dams do not compare to the injustices of racism in this

Emily Vogler
Set of cards designed for the workshops

country, but I see our failures to resolve such disparate challenges as being fundamentally connected. As the United States has become more politically polarized, it is increasingly difficult for people to come together for a civil exchange with those with differing beliefs and to find common ground. However, the complex social, economic, and environmental legacies that underlie the foundation and growth of the US will not just go away. They must be intentionally and carefully addressed at multiple scales, from the individual to the neighborhood to the national level. Whether we are working to repair the nation's dark legacy of centuries of racial inequality or to decide how to justly manage shared natural resources, we desperately need new methods to engage in civil discourse, tap into our shared humanity, and re-center decision making and civic action within local communities. New creative social practices have the potential to bring people together at the local level to compassionately discuss divergent viewpoints; to critically re-evaluate history, science, and personal values; and to fully consider the social and environmental repercussions of individual and community beliefs, identities, and actions. So, while the dam decision-making toolkit will hopefully help support communities making decisions about the future of aging dams, we hope it will also serve as a model for how to bring people together at the local level to discuss other difficult topics such as climate change, structural racism, and inequality.

For just as small streams eventually merge into mighty rivers, reconciling complex social and ecological legacies demands individual and community action that fosters an understanding of the interconnectedness of all communities—from upscale urban neighborhoods to downtown streets to rural towns to migrating fish, and all the humans and nonhumans whose lives intersect with them. ○

Notes

1. New England Sustainability Consortium, "The Future of Dams: Helping New England Communities Use Science to Make Decisions about Dams," newengland-sustainabilityconsortium.org/dams.
2. American Society of Civil Engineers, "2017 Infrastructure Report Card."
3. Moriah Balingit, Kayla Ruble, Steven Mufson, and Frances Stead Sellers, "Michigan Dam Disaster An Example of What Could Happen in Many Other Communities," Washington Post, May 22, 2020.
4. American Rivers, "Restoring Damaged Rivers," n.d.
5. Coleen A. Fox, Francis J. Magilligan, Christopher S. Sneddon, "You kill the dam, you are killing a part of me: Dam removal and the environmental politics of river restoration," Geoforum 70 (March 2016): 93. See also Sara E. Johnson and Brian E. Graber, "Enlisting the Social Sciences in Decisions about Dam Removal: The Application of Social Science Concepts and Principles to Public Decision Making about Whether to Keep or Remove Dams May Help Achieve Outcomes Leading to Sustainable Ecosystems and Other Goals in the Public Interest," BioScience, vol. 52, no. 8 (August 2002): 731.
6. Fox et al.
7. Quoted in Fox et al.
8. Fox et al., 94
9. Fox et al., 100
10. See Vandana Shiva, Water Wars: Privatization, Pollution and Profit (Cambridge, MA: South End Press, 2002) and Elinor Ostrom, Governing the Commons: The Evolution of Institutions for Collective Action (Cambridge University Press, 1990).
11. Forthcoming at damatlas.org.

→ Iona Fox   *Almanac, Week of September 13, 2017*

THIS WEEK i GRILLED DYLAN BECAUSE i THOUGHT iT WAS THE 25th ANNIVERSARY OF DiGGERS MiRTH COLLECTIVE FARM

DYLAN iS THE ONLY REMAINING ORIGINAL MEMBER OF THE COLLECTIVE — THE OTHER 2 CO-FOUNDERS SOON SWITCHED CAREERS. DYLAN KEPT FARMING, & HE KEPT THE COLLECTIVE OWNERSHiP MODEL.

HMM NOW THAT i THINK ABOUT iT iT'S THE 26th

SOOO

DiD YOU EVER HAVE TENSION OVER EVERYONE GETTING PAID THE SAME RATE

EVEN THOUGH SOME PEOPLE HAVE MORE EXPERIENCE?

NO...

WELL, THERE WERE A COUPLE YEARS EARLY ON WHERE iT FELT LIKE,   OH, iT'S JUST ME DECIDING WHAT TO DO,

& THEN TELLING PPL WHAT TO DO & THEY GO DO iT,

& WE SURVIVE & GET BY & THAT'S GOOD   BUT iT DiDN'T FEEL LIKE THE BENEFiTS OF A COLLECTIVE * YOU KNOW?

BUT iT CHANGED A LO-OO ACHOO!-

bzzz   ow ow   bzzz

SCRATCHING OUR HANDS IN THE CUCUMBERS

...PLUNGING THEM INTO POISON EGGPLANT POWDER

- CHANGED A LOT WHEN PEOPLE JOINED AND...

OR ELSE iT WAS JUST TURNING INTO LIKE

they HAD iDEAS, THEiR HEART WAS ALL THERE

...iT NEEDED PEOPLE W/ DiFFERENT PERSPECTIVES TO PUSH FOR THiNGS THEY CARED ABOUT

"DYLAN'S FARM... WHERE EVERYBODY GETS PAID THE SAME"

& i DON'T WANNA WORK ON MY FARM. i'VE GOT A LOT OF WEAKNESSES & i'M PSYCHED THAT PPL HAVE OTHER STRENGTHS

i'M STiLL LISTENING

...iT STARTED TO BE NICE

iT STARTED TO BE REALLY PLEASANT

THEN CAUTERiZING THE WOUNDS WiTH ACID TOMATO JUiCE

* time off!

Emily Haefner

PONDWEEDS

I floated out on the pond today
made of reeds, weeds, and swimming things.
My raft, a relic of land, steered itself toward far reaches.
Under trees,
into secret passageways of lichen covered arches.
Branches dipped through empty space to caress the water mirror.

I swung my bamboo paddle, seeking deeper eddies,
tangled in the jetsam.
I raced my other hand along the edge
fingers barely wet,
skimming.
I moved into the green thickets of water cities
ruled by regal damselflies and fire breathing dragonflies
who swarmed in patterns of blue and red.
Their wingbeats, too fast to see, turned a gold blur in the sultry noon sun.

As I lay above the water peering closer to the surface,
another face stared back—
a queen.
My straw hat was crowned with her water lilies;
flowers were woven in my golden curls
bouncing in rippling currents.
The reflection broke as a newt swam toward sunlight.
His inquiring face looked up to me:
Was I here to join the water world?
Revealing none of his secrets,
small webbed feet flickered as he turned away
beckoning me to explore the mysteries beyond.

A swish of the paddle,
hushed by the
soft harmonies of flying,
swimming,
living,
wild things.
I push off to the shores unknown.

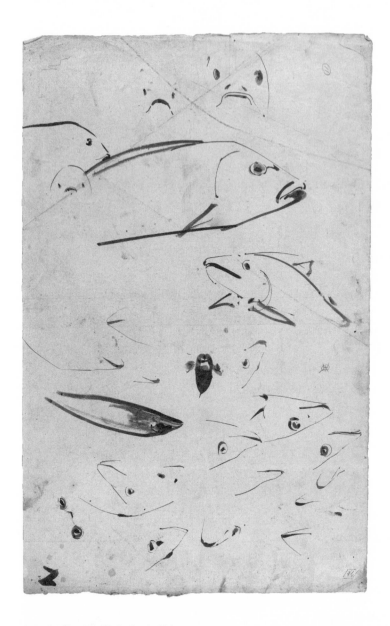

↑ **Gerrit Willem Dijsselhof** *Sketch of fish*
→ **NASA** *Toxic Algae Bloom in Lake Erie*, October 5, 2011

APRIL

Algae

Fish and Game, Department of Marine Resources

Austin Miles

REFUSING THE FARM

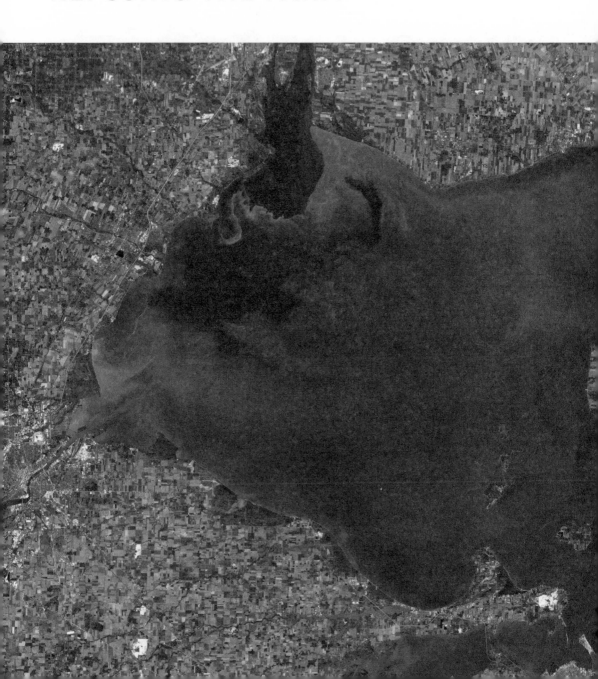

Northwest Ohio is a high-value wreckage, a swath of monocropped agricultural lands whose fertilizer runoff has precipitated an ongoing spate of harmful algal blooms in Lake Erie that began in the mid-1990s. The blooms are "harmful" because they are toxic—the species of cyanobacteria that composes the blooms produces a toxin called microcystin that is potentially lethal. This toxicity, combined with the fact that 11 million people depend on Lake Erie for drinking water, make the blooms a serious issue. Yet, despite the blooms' toxicity—despite having forced the Toledo city government to close its municipal water supply (the water supply for 500,000 people) for three days in August 2014, in an emergency known as the Toledo Water Crisis—lawmakers across the Lake Erie watershed have done little to resolve them.

The agriculture that causes the blooms may be traced to the initial erasure of Native lifeways in northwest Ohio through assimilation and removal, and the subsequent draining of the wetlands in northwest Ohio, an area white colonialists referred to as the Great Black Swamp. Prior to the imposition of settler-colonialism, Native peoples maintained reciprocal relationships among humans and the other-than-human.[1] Dana Bogart, in her history of the Great Black Swamp, calls this a practice of "'ecological knowledge,' which was systematic, relational, and interactional" with the land.[2]

In 1817, Michigan Governor Lewis Cass and General Duncan McArthur negotiated the Treaty of Fort Meigs with the Wyandot, Ottawa, Shawnee, and Seneca tribes, many of whom had been previously displaced by the Treaty of Greenville in 1795. With the 1817 Treaty, these tribes ceded their claims to roughly 4.5 million acres in what is now northwest Ohio. The Native claims remaining after the treaty were small, disconnected reservations drawn based on tribal lines and identification.[3]

What followed were programs of assimilation and white settlement, with the ultimate goal of removal. The federal government passed land use and farming policies to dispossess Native Americans through assimilation following the 1817 treaty. These policies had two important effects. First, they obligated the Wyandot, Ottawa, Shawnee, and Seneca tribes trying to maintain their claims to territories in the region to use the land "productively"—the ideal state of land according to US hegemony. Instead of maintaining a web of relationships, the relationship with land under this settler-colonial ideology was reduced to that of owner and property. Second, these policies encouraged white settlement in the region. As Bogart writes, "A region with Native Americans assimilated to Euro-American ways of life also made the land easier to sell to white settlers during this era." The state government also facilitated white encroachment on the Native reservations by funding the construction of infrastructure projects, such as roads and canals.[4]

The encroachment of white settlements on Native reservations in the region pressured Native peoples. The state and federal govern-

← **NASA** On September 26, 2017, the Operational Land Imager on the Landsat 8 satellite captured these natural-color images of a large phytoplankton bloom in western Lake Erie.

123

ments considered the Shawnee, Wyandot, Seneca, and Ottawa tribes as impediments to "civilization" despite their efforts to assimilate, as did the settlers. The Indian Removal Act, signed into law in 1830 by President Andrew Jackson, allowed the US government to orchestrate the removal of these tribes to lands in Kansas and Oklahoma. By 1843, autonomous Native reservations were gone from Ohio.[5]

The state government subsequently managed northwest Ohio to facilitate the draining of the wetlands—the last obstacle to "civilization" and economic prosperity in the region. Beginning in the late 1840s, Ohio enacted drainage laws like the Ditch Acts of 1854 and 1859, which funded the implementation of new drainage technologies such as clay drainage tiles. By 1920, the Great Black Swamp was completely drained and the conversion of northwest Ohio into a Euro-American economy of agricultural and industrial productivity was complete.[6] The US and Ohio had transmuted the land from an entangled composite of reciprocal relationships between humans and other-than-humans into capital—a simplified field of extraction. Draining the wetlands, which formerly would have filtered excess nutrients out of the water flowing into Lake Erie, produced and maintained the conditions

that eventually resulted in today's harmful algal blooms. These conditions are the traces of the unceasing settler-colonial project begun in the region in 1817—they form what some environmental humanities scholars term a "blasted landscape."[7]

I am interested in the possibilities latent in a blasted landscape. Even after a land suffers the worst destruction, I cannot accept the finality of calling the waste an apocalypse. Within these destructions lie potencies and the promise of regenerative futures. This hope for the future is akin to the emancipatory hope articulated by Jacques Derrida: "the attraction, invincible élan or affirmation of an unpredictable future-to-come (or even of a past-to-come-again)."[8]

Yet for the land that was formerly the Great Black Swamp, as well as Lake Erie, what emancipatory hope lies in industrial, mono-cropped agriculture? I am unmoved by the predominant solutions to harmful algal blooms, those that propose "best management practices" such as the use of cover crops or sustainable fertilizer use to mitigate the amount of nutrients flowing from agricultural land into the Lake Erie watershed. Practices like these serve as measures to make the settler-colonial project more palatable and do nothing to remediate the underlying violences enacted on the land. I am inspired instead by the Lake Erie Bill of Rights, which, in the words of the grassroots advocacy group Toledoans for Safe Water, recognizes rights "long violated and ignored." (The bill was voted into law by the residents of Toledo in February 2019, and struck down by a federal judge a year later, partly because the judge deemed that the bill granted the City of Toledo too much power).

Such a move poses questions about how to honor and enforce these rights. How might our relationship with Lake Erie and the surrounding lands emerge? What will they look like? As posed by the environmental humanities scholars Thom van Dooren, Eben Kirksey, and

Ursula Münster, "What forms of responsibility are required, and how might we learn to respond in other, perhaps better, ways to the communities taking form in 'blasted landscapes'?"[9] Granting an ecosystem rights is a move that presupposes a shift from the settler-colonial mode of inhabiting land—characterized by dispossession, erasure, enclosure, and extractive capital—to a mode of living that foregrounds the relationships that constitute the land. This mode is deeply informed by Indigenous relational thought.

The relationships composing land in Indigenous thought emerge from shared histories—they are processes of co-becoming, enmeshing all beings. What would it mean for land use in, for instance, the Maumee watershed, where the majority of nutrient input into Lake Erie originates, to restore these erased histories? Redress for the dispossessed Indigenous communities, certainly. As Lakota scholar Nick Estes wrote in *Jacobin*, "no society can ever have an ethical relationship to a place it stole." Atonement for this theft in the form of land repatriation and instituting new forms of relationship-attuned Indigenous governance would be a key part of decolonizing northwest Ohio.

The possibility arises that agriculture just should not be practiced in this region, or should, at the very least, be reduced. The ecologist

William J. Mitsch argues that 5 to 10 percent of the original extent of the Great Black Swamp should be restored in order to reduce nutrient flow into Lake Erie, which would entail the development of "wetlaculture" (a portmanteau of wetland and agriculture) approaches that intersperse wetlands with agriculture.[10] Tellingly, some varieties of wetlaculture are forbidden according to the Ditch Law of 1859, a law which helps maintains the settler conditions of drainage.[11] So there is an element of refusal in this solution. To refuse agriculture, because this agriculture depends upon ongoing destruction. In northwest Ohio, the path to decolonization requires (among other things) a complete dismantling of the structures that drain the wetlands. In the wake of this reflooding, a radically different agriculture could emerge, co-constituted by the shared histories of the wetland. An agriculture entangled in, as Dooren, Kirksey, and Münster write, "dense webs of lively exchange."[12] But the postdiluvian period might also cultivate an ethos, within that land, of non-production. A space left to the other-than-human. These sorts of decisions may result when we digest the history of a land as we reimagine its future—and if we let the land tell us how it wants us to proceed. ○

Notes

1. P.B. Burow, S.Brock, and M.R. Dove, "Unsettling the land: Indigeneity, ontology, and hybridity in settler colonialism," *Environment and Society* 9, no. 1 (September 2018): 57–74.
2. D. Bogart, "'My Great Terror, The Black Swamp': Northwest Ohio's Environmental Borderland," (master's thesis), (Miami University, Oxford, Ohio, 2015).
3. Bogart.
4. Bogart.
5. Bogart.
6. M.R. Kaatz, "The Black Swamp: a study in historical geography," *Annals of the Association of American Geographers* 45, no. 1 (March 1955): 1–35.
7. S.E. Kirksey, N. Shapiro, M. Brodine, "Hope in blasted landscapes," *Social Science Information* 52, no, 2 (May 2013): 228–256.
8. T. Van Dooren, E. Kirksey, E., and U. Münster, "Multispecies Studies Cultivating Arts of Attentiveness," *Environmental Humanities* 8, no. 1 (May 2016): 1–23.
9. Van Dooren et al.
10. W.J. Mitsch, "Solving Lake Erie's Harmful Algal Blooms by Restoring the Great Black Swamp in Ohio," *Ecological Engineering* 108 (November 2017): 406–413.
11. Mitch.
12. Van Dooren et al.

Alli Maloney *View of Cobscook Bay from Smithereen Farm, Passamaquoddy Land, 2019*

Lydia Lapporte

TOWARDS A SEAWEED COMMONS

All This Water

It gets slippery, reimagining terrestrial territoriality and property relations, thinking *with* water rather than of it. Ours is a small-scale platform of inquiry into the seaweed commons, formed in response to an uptick in seaweed aquaculture and the emergent legal status of rockweed in Maine. With our toes in the water, we look towards a politics of human and non-human flourishing at the intertidal. Weary of solutionism, technocracy, and enclosure in the marine biome, we inhabit a political-ecological framework that engages multiple scales of discipline and analysis. Our wild harvest model is part of an integrative farm ecology. As in the young farmer's movement, questions of scale, agency, and relation are key here—but now wet, salty, and immersed in a marine biome that spits back at us.

Grounding and Floating

At the low trophic level, seaweeds support the critters, oxygenate the waters, and constitute the foundational flourishing of intertidal places. Seaweedcommons.org relays: "Everything on earth grew from the sea. We are all beneficiaries of the blue-green algae (cyanobacteria) that invented photosynthesis. They lazed around in shallow waters converting the energy of sunlight into the sugars and carbon bonds of life, producing oxygen and ultimately our atmosphere."[1] Now algae growth, fresh or salty, acts as a material indicator of waters' happenings. Agricultural runoff springs toxic freshwater algae blooms while the warming and acidifying and otterless bays cause kelp forests to dwindle.

Seaweed is liminal; it resides at the threshold between wet and dry, sea and land. The shallow and sun-penetrated intertidal edges are good places for wading in, for learning to think and be otherwise.[2] To me, these buffer zones act as bridges: to oceanic relations, theoretical threads, the primordial soup of biological life. And these potent edges of life and wrack and scum might be particularly relevant now, as human and non-human relations erode and evolve at once.[3] Marine algal life resonates like some sort of language that needs listening to.

Caught

The seaweed caught me: floating and tangled somewhere between Atlantic rockweed and Pacific bull kelp. There's a boundlessness, a slippery, slimy joy of seaweeds—their rhythmic underwater dancing, translucent hardiness— agential and feminine and sensual. You might see it as muck, what painter May Stevens calls the "origin stuff"[4] of creativity: that which comes from within the thick water rather than transcending it. Withinness, forgoing the overhead perspective (state, god, bird) for one of submersion (ants, publics). Annie Albers writes about language and thread and the breakdown in her weaving:

In my case it was threads that caught me, really against my will. To work with threads seemed sissy to me. I wanted something to be conquered. But circumstances held me to

→ **Lydia Lapporte** *Situated*, mussel shell, paper, silk thread

threads and they won me over. I learned to listen to them and to speak their language ...
What I am trying to get across is that material
is a means of communication. That listening
to it, not dominating it, makes us truly active,
that is: to be active, be passive.[5]

This liveliness sounds to me like salt cakes
and nudibranchs and wrack, the yellow grey
green smell held in the fog. At low tide, matter
and language and being seem to slosh more
fluidly, diffusing terrestrially embedded
boundaries.

Thinking with Water

Feminist phenomenologist Astrida Neimanis
likes to describe the world as lived. She calls
water gestational, a medium of distributive

agency that helps us think across national,
species, and individual boundaries. As bodies
of water, we leak into each other. The physicality of hydrological cycles make it so: in drinking
and respirating and peeing, my water becomes
your water becomes non-human water. "Water
in this sense is facilitative, and directed
towards the becoming of other bodies."[6] Here,
a sense of constantly becoming other (non-
human, non-self) might trouble individualist
notions of discrete, bounded, and autonomous
subjectivities, responsible for our own health
and harm. Perceiving embodiment as wet
(porous, fluid, co-constituted) might flip this
logic, turning to an inherent corporeality,
collectivity, and porosity of the body. Accordingly, questions of human and non-human

selected text from Stacy Alaimo's
*Your Shell on Acid: anthropocene dissolves &
states of Suspension: trans-corporeality at sea*

This essay will focus on suspension as a ... of
buoyancy, a sense that the human is held, but not
held up, by invisible genealogies and a maelstrom of
often imperceptible substances that disclose
connections between humans and the sea. New
materialist and science studies that stress the
impossibility of an ontological divide between
nature and culture are invaluable
for contemplating seawater itself. The utter
unprecedented nature of the 'massive and rapid
changes caused by carbon dioxide emissions may be
best understood as a grand planetary experiment,
but an experiment that places the entire planet and its
creatures within the 'mangle of practice,' a
posthuman space where scientific practice is an
open-ended, reciprocally structured interplay of
human and nonhuman agency. We all live 'in the
thick of things' even though the everyday stance of
... and domination blocks this recognition.
... contemplate the ...

Lydia Lapporte *Selected Text*, paper stencil, salt, selected text by Stacy Alaimo

well-being connect more tangibly to politically and socially embedded structures of health and harm. Here, a tracing of waterways, bodily and otherwise, might make felt the material inevitability of our co-constitution with the world.

As sites for tidal attunement, for water retention and release, the seaweeds might nudge us towards an opacity and porosity of being. In its immediacy as as oxygen producer and ecosystem foundation, the seaweed is bearing a great deal of our collective knowledge and grief. Shimmering with boundlessness and generosity. Bathing plural life into being.

Water is a transitory element, and a "being dedicated to water is a being in flux."[7]

Water Carries

All this water. Melting, warming, rising, lapping into an increasingly oceanic future. The sea and waters that slope towards it become central in a moment of rapid biogeochemical change and political-economic erosion. A singularized understanding of ocean as romantic scape doesn't hold up. Rather, oceans might be plural containers for a number of contradictions: origin stories, uneven toxic distributions, dispossession (waters rise, home places are swallowed), plastic immortality, queer archives of feeling,[8] routes of violence and freedom, and vessels of memory.

Most of it—the runoff of a global political economy—winds up in the sea, and the lapping of the tides and swirling of the currents implicate and impact watersheds, bodies, and inheritances differentially. The violence is slow, looping, and intergenerational, produced and reproduced in material realities and bodies.[9] For instance, tidal swirls and current specificities render Arctic waters acutely concentrated zones of persistent organic pollutants and other industrial externalities.[10] Bioaccumulation up the food chain concentrates these afterlives in the fat of marine mammals and fishes. Through proximal physical and cultural relationship to these waters and marine animals, toxins are downloaded in the breast-milk of Arctic peoples, particularly Inuit women,[11] making it toxic to consume. Exposure to premature death is made material across lines of difference, both raced and sexed. The afterlives held by water are not just material, but historical and political, too. (See Christina Sharpe's *In The Wake: On Blackness and Being*, in which she interrogates the orthography of the wake of the slave ship in contemporary black life). In this way, water is a shared political conduit, holding, carrying, and iterating across temporalities and places.

Seeping, water resists neat resolve. A hydrocommons approach understands bodies of water as at once globally circulating and materially embedded within specific locales. Water prompts: "we are always different and always in common."[12] The inadequacy of this "we"—the we of abstracted and commodified water and the we of universalized humanist flattenings—reminds us that we are at once immediately situated and politicized bodies as well as entangled and diffuse in abstract pools. Here, I take up a material feminist critique of the too-easy "we" of humanity, while at once refusing the bounded "I" of humanism.[13]

Commons Posturing

Understandings of commons vary; ranging from communal land holdings to open-source knowledge platforms to creative housing arrangements. Something like a *quality of relations* among people and place, commons systems emerge as means to self-tended governance and use of shared resources. Relational and iterative, they hold negations and fluctuations in a nature-culture continuum. Commons are scaffolded by social infrastructures, some insurgent and subversive, others managerial and tedious. In robust commons systems, human and non-human flourishing is co-constitutive; we make the world and the world makes us back. Through the seaweed commons praxis, we are keen to think with commons frameworks as we negotiate the ecological and economic shape of seaweed use along Maine's coast.

Here, In Maine:
Rockweed On Rock On Rock On Rock

The biomass is here. The tides are big and boiling at twenty feet, and the lobstering rope frays like split ends. Depending on the light, rockweed is chartreuse like mustard flowers or olive or ropy deep-red. It's like black paint mixed with yellow, and there's often grey rock and fog, calm water in the bay. Sprawled out abundantly along Maine's raggedy coastline, rockweed makes Maine's intertidal zone—the place between low and high tide marks—legible.

Ancestor to all, rockweed is kin to more than one hundred fifty species: bird, bug, fish, bacteria. A rockweed bed is a support, a foundation, and a nursery for all teeny marine things and those in their wake.

In Maine, rockweed has traditionally been harvested within the public trust framework for personal use as fertilizer, food, and medicine. The public trust doctrine ensures public access in the intertidal zone for fishing, fowling, and navigation despite upland land ownership between the low and high tide marks. Thus, fishing, shellfishing, and worming operate within the public domain, while logging is subject to landed regulations that foreground private ownership.

As rockweed becomes an increasingly valuable resource for a wide variety of culinary, industrial, and cosmetic applications, unprecedented quantities of marine algae biomass are being harvested from the rocky coasts. Accordingly, its role in the marine environment and public trust are subject to exploitation by industrial harvesters. Acadian Seaplants rakes up boats of rockweed, dries and grinds, and then spits it back out to be sold as a "biostimulant" for golf course fertilizer and commercial seed coating.

The rockweed that blankets this coastline lives at a convergence of ecological, political, and economic entanglements: legacies of extractive industry and the social wounds that follow; potent marine assemblages; corporate speculation; rural disenfranchisement and dispossession; landowners only here to summer; fisheries drama and depletion; tourism economies; and lobster monoculture. As commercial harvesters continue to encroach rockweed beds, the value of the cut rockweed settles not here with the laborers or the onlookers or coastal sediment, but away. The blue harvesting boats come and go, ensconced in the watery milieu that forecloses easy regulation.

Conservation, Enclosure

Concerned with the lack of regulatory infrastructure in the nearshore zone, conservationists spent years advocating for biological conservation frameworks like no-cut zones in rockweed beds of high ecological import. Little came of this, and industrial rockweed harvest increased four-fold between 2001 and 2017.[14] Frustrated and somewhat desperate, shore-proximate landowners and conservationists imagined effective rockweed protection as private and aimed to prevent rockweed's overharvest by securing ownership claims to it. They reasoned that privatization would require landowner permission prior to harvest, which would in turn reduce industrial access. Their claims to private ownership were bound by rockweed's legal status as either a fishery in the intertidal or landed organism. The question became: is rockweed fish-like and publicly accessible for cutting, or tree-like and privately held, subject to landowner control?

In 2019, the Maine Supreme Court sided unanimously with landowners in *Ross v. Acadian Seaplants*, holding that rockweed is private property that belongs to the adjoining upland landowner who owns the intertidal soil.[15] The decision functions as a partial win for rockweed conservationists, theoretically

setting up barriers to industrial overharvest. But there is a disconnect between the new enforcement policy and what is happening on the water, all under the discretion of the Maine Department of Marine Resources (DMR) whose allegiances do not tend towards an "ancestor-algae" framework of reverence. Despite the ruling, Acadian Seaplants continues to cut seaweed in conservation areas without permission. Now, landowners can call marine patrol if they witness unconstituted rockweed harvest proximate to their land, but there is no effective enforcement action by the DMR.

Civil litigation is not a fine-tuned tool for enforcing this kind of industrial theft, a structural trouble with the conservationists' method of resistance to rockweed extraction. In practice, this privatization-as-conservation regime sets a crude and somewhat empty precedent for relating to the seaweed community. By granting individual landowners regulatory faculty around seaweed harvesting, the court's decision reinforces individualistic, linear, and exclusive conservation regimes that have the effect of exacerbating class tensions. It aligns intertidal regulation with private property ownership rather than creating opportunity for collective, mutually agreed upon, and enforced systems of community use frameworks. Something akin to the doctrine of *aqua nullius*, or a logic of ecological citizenship as contingent upon private land and shore holdings.[16]

Rockweed Commons

We intertidal stakeholders peering out from the shore fumble along without much data, without much time to work through our habits of mind and of use. In response to the current methods of resistance, we aim to imbue the rockweed discussion with the complexity, nuance, and creativity of a young farmer perspective. It takes eleven to sixteen years for a four-foot tall organism to recover from being cut sixteen inches from the holdfast.[17] How might we construct seaweed economies that consider rockweed time? Rather than imagining rockweed worlds as entirely separate from human ones, in need of paternalistic protection, what if we asked: How could the harvest of a public trust–dwelling organism center common systems of knowledge, control, and value? Rather than operating from the mindset of scarcity economics, what if we moved towards a vision of transformative and restorative—rather than retributive—environmental work, grounded in flourishing rural economies?

We could begin with the emergence of formal territoriality or sector allocation to ensure that each seaweed ledge is harvested only once during a season. We could advance citizen-science data collecting initiatives, social reimaginings of property relations, deeper understandings of social/ecological histories at multiple scales, and greater transparency and coordination between harvesters. We're trying to see, wading through incursion from a distance, treading the highly mediated modes of knowledge production and distribution that characterize the slow violence of anthropocene data visualization. We're trying to imbue another perspective—one from below, or from the edge—into the public discourse. Insurrection and vibrancy for and from our places.

There's much more work to be done here, at the wet dry edges.

Aquaculture
As the Gulf of Maine warms rapidly—more so than all but three other water bodies— we experience a future-tense perspective of

oceanic space. More acid, more sporadic fresh, more hot, a temporal and biogeochemical unmooring. As coastal citizens, an institution, and as humans in the anthropocene, we're learning to swim in new waters. We turn tentatively towards seaweed and shellfish aquaculture, farming in the sea, so as to not strain the wild marine commons that we harvest: alaria and nori and dulse and digitata and sugar kelp and sea lettuce. Ocean gardens and integrated multi-trophic aquaculture schemes sound rosy, but we sense potential for extractive tendencies as well as systems of mutual flourishing within the emerging sector.

Now, eight thousand fingernail-like baby oysters pass seawater through their bodies, filtering muck and phytoplankton in our four hundred square-foot Limited Purchase Aquaculture (LPA) zone at Schooner's Cove. We're fumbling, joyfully. Making a path through materials design from algicide-laden and man-centric cages to cork buoys and waxed linen rope. We aim to move towards an aquaculture in Maine's abundant waters that will center sustained, appropriately scaled rural development. The time is now to think ahead, to scaffold the institutions, to get to being and learning with the humans and non-humans of the intertidal. In thinking and acting together: commons, water, ocean, seaweed, and farmer.

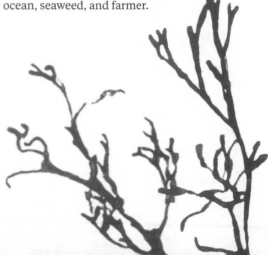

JOIN US!

Through a commons-informed approach, Seaweed Commons, a project of the Green-horns, seeks to form ecologically minded seaweed coalitions, support and inform public discourse, and support algal literacy in service to an appropriately scaled, just seaweed economy. In orienting toward more socially livable reimaginings of ownership, we center entanglement between social discourse and environmental matters. We create citizen science and public engagement platforms to better understand ecology and policy consider-ations that impact our coastal communities. Through our tentacled projects, we leverage art, film, public forums, research, and online informational resources with critical joy, so as to not overwhelm the lively ecology at the threshold of land and sea.

Our projects include:

* Annual Seaweed Symposium in Pem-broke, Maine
* Media production: a film, webisode, and podcasts
* Applied research into alternatives to plastic in aquaculture: articulating and experimenting the "ocean garden" model
* The research and implementation of a "best practices" certification for farmed and wild-harvested seaweed products
* Marine-commons art exhibit at the Pennamaquan Gallery in Pembroke: summer 2021
* Three Limited Purchase Aquaculture zones to tend the wild-domestic oysters, seaweeds, and the waters ○

Notes

1. Severine Fleming and Lydia Lapporte, "Evolutionary History of Marine Algae," seaweedcommons.org, Spring 2019.
2. Donna Harraway, *Staying with the Trouble* (Durham, North Carolina: Duke University Press, 2016).
3. See the work of Vijaya Nagarajan and Terry Tempest Williams.
4. Macarena Gómez-Barris, *The Extractive Zone: Social Ecologies and Decolonial Perspectives* (Durham, North Carolina: Duke University Press, 2017), xiii. Gómez-Barris describes an extractivist logic in which "the material and affec-tive production of extractive capitalism crushes vernacular life and its embod-iment. Extractivism assaults peripheral spaces, inflicting uneven pain across regions where Indigenous majority com-munities continue to organize life and proliferate it, even in sites of extreme pressure and violence."
5. Annie Albers, "Material as Metaphor," *Selected Writings*, The Josef & Anni Albers Foundation, February 25, 1982.
6. Astrida Neimanis, "Introduction: Figur-ing Bodies of Water," *Bodies of Water: Posthuman Feminist Phenomenology* (New York: Bloomsbury Academic, 2017).
7. "On Brathwaite's Tidalectics," NTS, June-November 2017.
8. See Neimanis.
9. See Rob Nixon, *Slow Violence and the Environmentalism of the Poor* (Cambridge, Massachusetts: Harvard University Press, 2013). Nixon emphasizes the constancy of environmental racism in his description of the slow violence of climate change: "a violence that occurs gradually and out of sight, a violence of delayed de-struction that is dispersed across time and space, an attritional violence that is typically not viewed as violence at all."
10. "The Mother's Milk Project of St. Regis," *sexgendernature: understanding ecofeminism in popular culture*, March 20, 2011.
11. "The Mother's Milk Project of St. Regis."
12. Neimanis.
13. Rosi Braidotti, *The Posthuman*, quoted in Neimanis, 63.
14. Ben Goldfarb, "A Fish Called Rockweed," *Hakai Magazine*, May 29, 2018.
15. Maine Supreme Judicial Court, *Kenneth W. Ross et al. v. Acadian Seaplants, LDT.* Decided March 28, 2019.
16. Brenna Bhandar, "Chapter 1: Use," *Co-lonial Lives of Property: Law, Land, and Racial Regimes of Ownership* (Durham, North Carolina: Duke University Press, 2018).
17. Robin Hadlock Seeley, "Seaweed Values and Protection/Management International Experience," *coastwatch. org*, May 2016.

Jacob Miller and Matthew Sanderson

WHAT WE NEED TO DO TO CELEBRATE THE 100TH EARTH DAY

Using COVID-19 to Root Down

Our species is at a historic crossroads. In December 2020, we find ourselves nearly one hundred years past the Spanish Flu pandemic, ninety years from the onset of the Great Depression and the Dust Bowl, seventy years since Aldo Leopold's "Land Ethic," fifty years since the Nobel Peace Prize was awarded to Norman Borlaug for sparking the Green Revolution, and just as much time since the inaugural celebration of Earth Day. Although the Green Revolution's contributions have provided extraordinary advantages, their unintended effects have also helped deepen humanity's misalignment with the natural world. This misalignment created the conditions for the COVID-19 pandemic, which overshadowed this landmark anniversary, preventing us from pausing to reflect on its significance.

What was already clear to many, and has become clear to many more, is that the entrenched approaches of the food systems that were disrupted by the pandemic were not "normal," so the last thing our society should aim to do is to "get back to normal." With this in mind, let's immerse ourselves in the long view. To answer "where do we want to be in fifty years?" we must first explore where we are: a time with signals all around us. The world is sick from anthropogenic encroachment upon its ecosystems, and it is telling us so.

Human encroachment on ecosystems destroys habitats, increases stress and disease vulnerability, and reduces biodiversity.[1] Genetic analysis of SARS-COV-2 suggests that COVID-19 originated from a bat,[2] likely spreading the virus through an intermediate animal, which, as a result of urban sprawl, humans encountered. There is precedent for coronavirus outbreaks that follow this pattern of infection: SARS, for instance, spread from bats to cats to humans. More broadly, the spread of infectious diseases among wildlife and the zoonotic transfer of primordial and novel pathogens to humans has occurred before; the bubonic plague and HIV are infamous examples. Human encroachment renders complex, adaptive ecosystems less resilient to future shocks.

Repurposing land for urban development not only maintains but deepens our reliance on agriculture. And the dominant approach has been industrial. Its proponents have long capitalized on assembly-line efficiencies to swallow competition, raze forests, destroy habitats, and plant monoculture acres that raise risks of human exposure to novel diseases. COVID-19 is yet another plot point in a series of viruses that

Robert Rauschenberg *Earth Day, April 22 Poster*, 1970   Poster for the benefit of the American Environment Foundation, from an edition of 10,300 of which 300 were signed. Published by Castelli Graphics, New York   © Robert Rauschenberg Foundation

have swept the world since 2002: SARS, H5N1 (bird flu), H1N1 (swine flu), MERS, and now COVID-19.

This pattern is not only alarming, but exhausting. As Dr. Rob Wallace, an evolutionary biologist and leading expert on the spread of influenza and other pathogens, has explained, "Many epidemiologists have a certain exhaustion coming out of coronavirus if only because we've seen this multiple times, and every effort to raise the alarm to say, 'Hey, here we need to change our agricultural practices'... has been largely ignored—in part because [we are] pretending that we can continue to cut into the forests and grow species that select for the most dangerous pathogens, and not suffer the consequences."[3]

COVID-19 has made us aware of the consequences. In the United States, the supply chains strained and caused scarce demand, forcing farmers to dump mass quantities of produce and milk. Grocery stores struggled to maintain reliable stocks and used "military-style" distribution tactics to ensure stability in supply chains.[4] More Americans became unemployed, faced empty store shelves and price gouging, and questioned how to feed themselves. Our commodified food system, consolidated and centralized as it is into precious few processors, distributors, and stores, was revealed to be vulnerable.

The US government struggled to cease the hemorrhaging. In the first half of 2020 alone, the American Farm Bureau pleaded with the USDA to "rescue all sectors of agriculture"[5] while multinational corporate profiteering via concentration of market power raised investigations into their price fixing.[6] President Trump invoked the Defense Production Act, which in turn allowed large meat processing plants to continue to operate without the worry of being held liable for the safety of their workers.[7]

Now, back to our imagination. It is 2070. COVID-19 ceased to be the topic of everyday conversation; it now appears in online textbooks as one example in a long list of pandemics. What has changed? The answer to that question depends on how we answer another: What *needs* to change if history is to reflect kindly on us?

If industrial agriculture lies at the genesis of so much of our sickness, the path to a healthier world is clear: We must acknowledge limits while (re-)learning to appreciate the diversity that remains in our own places. This entails "rooting down," or reconnecting and recommitting to place, to a diversity of places and resilient ways of living. An example of what this could look like: we both grew up on acres in family farms and rural communities, learning valuable lessons about the value of diversity for resilience. We calculated how necessary a thirty-minute trip into town would be. We could go long stretches without going in for groceries because there was enough chicken, beef, vegetables, and fruit preserves in the freezer to sustain our family. We used organic fertilizer and a lot of hard work. We paid more attention to farmers markets and church gatherings than stocks and futures. We relied on a symbiotic relationship with neighbors to fix fences, swap equipment, share produce, and trade chores.

It is easy to say "do as we have done." It is much, much harder to do what should be done. How many minutes in a day do we really spend thinking about what life will be like in 2070? Of those minutes, how many are spent thinking about an individual lineage rather than the state of the culture? At the heart of the necessary transformation are perhaps the greatest challenges facing our species—thinking beyond ourselves, thinking about future generations, and thinking for natural ecosystems. Although these may seem impossible, our hope stems from an ardent belief in the word "necessity." Necessity will force us to choose the right path.

The path will connect land, people, and community at the scale needed to sustain each. Although there is no one scale for these relations, a proper farm scale maintains an "eyes-to-acres" ratio, as Wes Jackson and Wendell Berry put it, where there are enough people on the land to work it without reliance on excessive fossil fuel inputs and technological subsistence.[8] This is not the only approach, especially since, among other access barriers, land access is a privilege few have, ourselves included. There is simply not enough land, even if distributed equally. By our rough estimate,[9] if you divided up all the terrestrial acres on Earth (not all of them habitable) among the current population, each person would receive just under five acres. Still, small-scale farms are a necessary part of the solution.

In sum, in 2020 we lived with the consequences of our consolidated, encroaching, out-of-scale, and monocultural food system. By 2070 the two of us will probably be dead. But our wills will demand that our kin read this piece on the 100th Earth Day. If they do, they will realize one of two things: either the food system continued down the same path, sickness prevailed, encroachment-related viruses became commonplace, and articles like these ceased to do any good. *Or,* our hope came true, and those who survive us live in rootedness, driven by the daily dictates of the land as their agrarian ancestors were once driven. Presence, community, and connectedness return.

We now speak directly to our kin: If the former, we deeply apologize. We did what we could, but it was not enough. If the latter, we rest in peace knowing that you are on the right path. In your journey we wish you safety. We wish you peace. We wish you healing, and we wish you land. ○

Notes

1. J.R. Rohr, C.B. Barrett, D.J. Civitello, M.E Craft, B. Delius, G.A. DeLeo, P.J. Hudson, N. Jouanard, K.H. Nguyen, R.S. Ostfeld, J.V. Remais, G. Riveau, S.H. Sokolow, D. Tilman, "Emerging human infectious diseases and the links to global food production," *Nature Sustainability* no. 2 (June 2019): 445–456.

2. M.F. Boni, P. Lemey, X. Jiang, T. Tsan-Yuk Lam, B.W. Perry, T.A. Castoe, A. Rambaut, D.L. Robertson, "Evolutionary origins of the SARS-CoV-2 sarbecovirus lineage responsible for the COVID-19 pandemic," *Nature Microbiology*, July 28, 2020.

3. Stephanie Bastek, "How Global Agriculture Grew a Pandemic," *Smarty Pants*, no. 120, March 13, 2020.

4. Kathy Gilsinan, "A Marine General's Next Battle: Grocery-Store Logistics: Larry Nicholson once led 20,000 troops in Afghanistan; now he's making sure you don't run out of food during the coronavirus crisis," *The Atlantic*, March 25, 2020.

5. Philip Brasher, "Farm Bureau makes sweeping appeal to USDA to rescue producers nationwide," *Agri-Pulse*, April 4, 2020.

6. L. Nylen, L. Crampton, "'Something isn't right': U.S. probes soaring beef prices," *Politico*, May 25, 2020.

7. J. Jacobs, L. Mulvany, "Trump orders meat plants to stay open in move unions slam," *Bloomberg News*, April 28, 2020.

8. Sasha Stashwick, "More 'eyes to acres'— and grocery aisles, dinner plates, and lunchboxes." *NRDC*, October 3, 2012.

9. The Earth's solid surface is 57,500,000 square miles. There are 640 acres per square mile, meaning there are 36,800,000,000 total acres on land to be divided up. The US Census Bureau's World Population Clock, as of August 27, 2020, estimates 7,676,000,012 people. Dividing acres of land by number of people results in 4.79 acres per person. This assumes all the acres are habitable, livable, accessible, and could be divided up or redistributed, which is nowhere near true.

An ATLAS Arts commission
by Cooking Sections

CLIMAVORE

On Tidal Zones

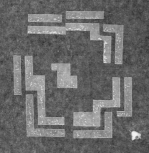

Set on the intertidal zone at Bayfield, in Scotland, *CLIMAVORE*: *On Tidal Zones* explores the environmental effects of aquaculture and reacts to the changing shores of Portree. Each day at low tide, the installation emerges above the sea and functions as a dining table for humans, with free tastings of recipes featuring ocean purifiers: seaweeds, oysters, clams, and mussels. At high tide, the installation works as an underwater oyster table. The oyster table is activated by Cooking Sections in collaboration with local stakeholders, residents, politicians, and researchers. The long-term project aims to look at forms of eating that address environmental regeneration and promote more responsive aquacultures.

Cooking Sections has teamed up with restaurants and food establishments across the island to incorporate *CLIMAVORE* dishes in their menus, featuring seaweeds and bivalves. These plants and animals purify the water by breathing, and can also help monitor the improvement of water quality along the coast of the island. ○

Anamarija Frankić

WHAT ARE WE LEARNING FROM OYSTERS?

Restoring with Nature

Picture oysters piling up to seven-meter high reefs and spreading thousands of acres in area, building 3-D structured habitats in every sea and along every coast of every continent. Due to the dearth of existing oyster reefs, it is hard to imagine vast oyster populations or understand their benefits to our global ecosystems. Let me share a few incredible facts about this bivalve shellfish which over the course of my marine science education and research has changed my life and opened my eyes to the wisdom of nature around me.

Oysters have diversified into hundreds of species adapted to specific coastal marine and estuarine environments, becoming one of the world's most amazing coastal engineers. Evolving for five hundred million years, this intrepid shellfish has shifted with temperature change and many sea level rises, catching up and adapting to natural changes through the Ice Age, until the Anthropocene era. We know today that one adult oyster can filter between one to two hundred litres of water per day, approximately fifty gallons. Their role in nature is of great importance to biodiversity, ecosystem health, and resilience.

Unfortunately, global oyster populations and their reef habitats have been reduced by approximately 85 percent in many coastal ecoregions and by 99 percent in most bay areas (estuaries). Declines in oyster reef population are mainly due to anthropogenic activities that lead to over-harvesting, habitat destruction, coastal water pollution, and disease spread. When the harmony of a natural system is harmed, it is more prone to diseases. However, because of the aquaculture industry, oysters continue to appear on restaurant menus around the world. This amazing, healthy delicacy is an important food source for humans, a keystone species that is missing from one of the most important niches in nature—coastal and marine systems.

Oyster reef designs are formed through the gregarious settlement of oysters, preferably on their own shells or any other calcium carbonate substrate. They create three-dimensional structures promoting and supporting many important ecological services. As ecosystem engineers, oyster reefs help facilitate biodiversity across ecosystems, enhance benthic and pelagic coupling by harboring juvenile and smaller fish species, create natural coastal buffer zones absorbing wave energy, and prevent erosion. Oysters filter water to remove nitrogen bound in phytoplankton and organic particulates, as well as promoting biodeposition and bioremediation. Their shells have the capacity to buffer, maintaining marine and estuarine hydrogen ion levels, measured as pH levels. We should not undermine their ability to reverse present water pollution and ocean acidification, although it is hard to visualize the historic abundance of their presence in the global marine environment because we don't have a single example of a healthy oyster reef to learn from.

There is not enough space here to describe everything we have been learning from oysters and what else they can do and teach us, but growing numbers of people are aware of their wisdom and benefits. Throughout their lifespan, oysters help reduce turbidity and improve photosynthesis in deeper waters, promoting the growth of submerged aquatic vegetation (SAV). The symbiotic relationship between deepwater oyster habitats and SAV supports synergistic ecosystem benefits including sediment stabilization, habitat creation, and improved water quality. Along coastlands, oysters prefer the vicinity of salt marshes. These three keystone coastal systems (salt marsh, oyster reef, and SAV) act in unison to create some of the most biologically productive areas and promote overall ecosystem health and function in coastal systems and estuaries. Natural systems that love to work together with oyster reefs vary globally and include mangroves, coral reefs, rocky shorelines, kelp, mud flats, sandy beaches, and dunes. Essentially, every natural coastal and marine system requires filter-feeders and healthy waters.

Considering the beauty of nature's designs, which create the most efficient possible collaboration between systems by providing the nexus of water, energy, and food between land and water, I and other researchers have asked, How would nature restore oysters and the reefs?

Working with Nature: Biomimetic Restoration
Oyster reef restoration projects have focused on reversing the trend of inhibited areas rather than on rebuilding self-sustaining reefs to promote ecosystem health, functions, and services. Oyster populations are dynamic and have the capacity to carry out ecological services in any given location, depending on that location's environmental conditions. Biomimicking natural solutions is the most environmentally, economically, and socially justified approach. Therefore, my work in reef restoration has been based on biomimicry principles that are non-negotiable and involve the creation and support of conditions that are essential not only for life itself, but for any

ethical, moral, and environmentally sound activity, design, or decision.

My experience in restoration is based on twenty-five years working mainly in the United States, specifically with *Crassostrea virginica* (eastern oyster). To help this species, we had to provide hard benthic substrates (cultch) on which naturally present oyster larvae could settle. We were able to use this method with restoration projects in Wellfleet Harbor and Chesapeake Bay. However, in the areas with limited or no existing oyster population—Boston Harbor, Nantucket Island, and Orleans Harbor, for example—restoration was supported by the "spat on shell" propagation approach. In mucky and soft sediments, where benthic substrate was unsuitable, we used floating structures made from materials such as recyclable plastic and green cement. The environmentally friendly materials were based on Green Chemistry technologies and identified to be the most suitable to support oyster reef restoration, including various green cements manufactured by BluePlanet Ltd, ECOncrete, and Grow Oyster Reefs, LLC.

How many oysters do you think are necessary to support a healthy population and a reef habitat? In 1960, native reefs consisted of 5,895 oysters per square yard.[1] We used this number to visualize and plan the vast potential of oysters in restored areas. In the Wellfleet Harbor project, we established a two-acre pilot site and in two years restored six million oysters by placing tons of recycled shells for oyster settlement. Restored oyster populations improved the harbor's local water quality and biodiversity, including fish and crustacean species, as well as improved sediment quality and shoreline protection.[2]

A biomimicry-based approach to oyster habitat restoration seeks to ensure that infrastructures and activities are based on six core criteria:

1. Does the newly restored habitat (structure) have the capacity to evolve and survive? The goal is for the restored or new communities to become self-sustainable in situ, evolve naturally, and provide ecosystem services and functions.

2. Is the restoration project resource efficient? The multifunctional design in oyster reef restoration relies on the nexus of water-energy-food, a reciprocal process established between the building structure, surrounding water, benthic communities, and other degraded coastal habitats like salt marsh and kelp.

3. How does the restored site adapt to changing conditions and model resilience? Habitat and ecosystem resilience depends on their interconnectedness and interdependence as living things to adapt and evolve locally without wasting energy or materials. Collaboration, not competition, is the driving force in evolution.

4. Are development and growth integrated in the restoration activity? Structural designs of oyster habitats are modular and nested within water columns and benthos; over time, they integrate development and growth to self-organize, developing complex food webs and trophic interactions.

5. How is the restoration locally attuned and responsive? The complex ecological functions provided by an oyster reef at a restoration site must be determined and incorporated in designing the built structures.

6. Does the restoration use life-friendly materials, water-based chemistry, and self-assembly? This principle requires that structures are built using local, natural,

untreated materials—materials that are degradable, sustainable, recyclable, and not harmful to water. In short, building materials must support conditions conducive to life.

Healthy water quality is the most important factor supporting coastal habitats' resilience, function, and biodiversity. There are numerous missing ecosystem services that oysters and their reefs can provide in polluted and degraded marine ecosystems, including bioremediation.

What we—I, my students, kids, elementary and high school participants, and local communities—have learned from working with oysters is that they are able to hear and feel us. The more love and songs we share with them, the more they thrive. Years ago, we were not able to scientifically prove this impact, but with today's new technologies able to capture electromagnetic waves and bioelectricity, we can monitor and measure the wavelengths and communication between organisms in water and air. Nature can "hear" us, always; all the world's species communicate all the time. It is on us now that we listen, even if we cannot hear them.

Native Oyster Reef Restoration Ireland (NORRI)
In Ireland, we have initiated the NORRI project[3] to restore *Ostrea edulis*, an endangered native European flat oyster. Historically, this oyster species has embraced and provided numerous ecological services and functions along the east and southeast coastline of Ireland, establishing eighty kilometers of extensive reefs from Wicklow Head to Ravens Point. Arklow was the main port for oyster fisheries in the 1800s, with a harvest of forty million oysters in 1863. Today, the whole of Ireland lands just over two million native oysters per year. Every single harbor

in the world has a similar story. The oyster reefs are gone but remind us of their presence with accumulated oyster shells along the beaches.

Together with the local community, we started the initiative to restore their historic beds and established Native Oyster Reef Restoration Ireland CLG (NORRI.ie). We have selected two suitable sites for oyster reef restoration along Wicklow County's coastal marine area, and the main goal is to establish two no-take areas with Biomimicry LivingLabs and to learn and share by doing the restoration. In collaboration with NORRI, Wicklow County Council hopes to reintroduce this species and restore oyster reef areas so that Wicklow and neighboring counties can once again

Photos by **Anamarija Frankić**

benefit from the presence of this native, keystone species in the Irish Sea.

People have a tendency to forget what is beneath the waves, below the surface of the sea. A vast amount of earth's territory is underwater and holds promise for great ecological and economic health and wealth—if managed in an environmentally responsible and sustainable manner.

What can you do? First, say thank you to oysters when you eat them next. Second, please bring shells back to a recycling bin or back to the sea where they belong. One in a million oyster larvae might find a home to settle on them. ○

Notes

1. G.R. Lunz Jr., "Intertidal Oysters," *Wards National Science Bulletin 31*, no. 1 (1960): 3-7.
2. Anamarija Frankić and Curt Felix, "Oyster Reef Restoration Project 2012/USDA Report," 2012.
3. Our documentary, *Deepwater - NORRI Project*, is available on YouTube.

↑ **Anna Atkins** *Dictyota atomaria*
→ **Dorothea Lange** *Pea pickers line up on edge of field at weigh scale, near Calipatria, Imperial Valley, California*

Ceremony •
Health

Phoebe Paterson de Heer

THE BODIES THAT WORK IT

Illness, Farming, and Access

I didn't know it would be my last day at the farm. It was just another Wednesday harvest in early autumn, cool and sunny. Tasks I knew well unfolded ahead: plucking cherry tomatoes from tall vines in the greenhouse, forking the soil in the carrot bed to loosen it, slicing the floppy leaves off a hundred leeks.

This day, though, there was a wrongness that I hadn't felt before. I felt harassed by the sun—flushed, light-headed, and weak. My limbs bellowed for rest. My mind was dull and slow, dripping with fatigue. I groaned in the field. Though I wanted to, I did not stop.

When I came to this work two and a half years earlier, I'd brought my illness, my compromised and beleaguered body. I have endometriosis, a chronic disease that makes bits of uterine lining grow and bleed in places where they shouldn't. I was often in pain—but farming, I found, brought me joy and meaning and community. I hurt anyway, so why not do the thing I loved?

When I came to this work, I was told it would take six months for my body to adapt to the labor, for my limbs and back to gain the strength and stamina to not feel sore at the end of every day of hoeing, hauling, and crouching. At first, I felt emboldened and capable. This land was now my livelihood, this work my exquisite challenge.

Six months? I thought. *Not even.*

As it happened, six months almost broke me. I was taking days off nearly every week. On mornings that first winter, I would get out of bed, put on my farm clothes, make a cup of tea. I would sit on the couch in a dark room and crumple with the pain and exhaustion of a body worn thin.

I moved to farming part-time. I ignored the voice that told me this was a kind of failure. And for a long while, it worked. I could still call myself a farmer even though I had to find other, less strenuous jobs to support myself. On farming days I worked as hard as I ever had, and on non-farming days I recovered.

Over scraping hoes, bubbling salad leaves, and sweaty lunches, my boss and I had many passionate conversations—about why farming was important, why we should be doing it, why this vocation could be radical and a form of both protest and world-bettering. We talked, too, about the work's toll on our bodies. Some afternoons when my uterus was screeching at me, my boss told me gently to go home. Increasingly he was taking days off himself, seeing a physiotherapist for his knees, his back, his neck stiffened by hours on the tractor.

What we never did, though, was talk about this pain and bodily wear as a threat to our devotion to farming, a threat to our idea of it as a

Jean-François Millet *The Reaper*

radical vocation. We didn't discuss accessibility and how able-bodied people who grow food for their communities are often not able-bodied forever.

I remember our exasperation. Exercise is supposed to be good for you, so why was this labor causing us so much pain? We ignored the possibility that this work—high impact, repetitive, hour on hour—was not quite the work our bodies evolved to do. And I, at least, ignored the signs my body gave me. So after two and a half years, that Wednesday harvest turned out to be my last. I went home and crashed, and when I woke again I was not the same. I would not be the same for over a year.

Slowly, the long haze of chronic fatigue began to clear. As I painstakingly extended my daily activity from a walk to the front door to a walk around the block, then a trip to the supermarket, I had a lot of time to think about what had gone wrong.

Farming supported my health in many ways. It fed me incomparably nutritious vegetables. It got me outside, in the air; it kept my hands in the earth, my body moving. Still, I wonder, can farming really be healthy if it harms us, too? If there is no place for me on the land anymore, with my broken body, who else is being left out?

I want to imagine another way of working. I want to believe that growing food doesn't have to mean pain and exhaustion and long days. I want to be part of a land-based community that supports the health of all its members—by paying fair prices for good food, by spending time with growers and learning their burden as well as their joy. I imagine a community that agitates in thanks for the well-being of its farmers. One that understands, deeply, that the healing of land cannot take place without devoted attention to the bodies that work it. ○

Bernadette DiPietro

LAUNDRY LINES

I was raised in an extended Italian family in the gray shadow of a Pittsburgh steel mill that bordered our property. Laundry was a daily chore; we hung the wash in all kinds of weather. It was a task that had rhythm and order. It brought us close to the earth, being grateful for a sun-filled day and the ability to feel the earth beneath our feet. Wringer washers, clothespins, balls of white cotton twine, and wicker baskets covered with floral oilcloth were everyday words. It was there that the arrangements of the clothes, their color relationships, and their movement captured me.

In 1958 I won my first camera by selling subscriptions to magazines. It was lime green and had a boxy, plastic shape. I remember the feel of the camera in my hands and the sound of the clicking shutter. It was then, at age eleven, that I began to develop my eye for composition. I still love looking through that small opening and creating visual arrangements.

In 1975 I received my first wringer washer as a gift on Mother's Day. Shortly after, I began photographing clothes hanging out to dry as my own personal journey, and in 1978, I began traveling to other countries and photographing

Photos by **Bernadette DiPietro**   *Da Nang, Vietnam*

Wash Day for Dolls

foreign clothes hanging out to dry. Colorful clotheslines are much easier to find abroad. I was inspired by early childhood memories of my Italian mother, who set me on this voyage.

In 2000 I purchased my first, used, electric automatic dryer at age fifty-three, and in 2002 I bought my first brand-new dryer, although I still prefer hanging clothes out to dry. I like the dance with nature and the experience of the outdoors when I'm hanging clothes. I like being aware of the sky and the blowing wind. I like the smell. I like the color, design, and feel of weathered, wooden clothespins. I like creating arrangements of colors, textures, and shapes against the landscape. It's my picture that I paint against the sky—an instinctive, moving sculpture.

In 2003 my mom passed away at ninety-one, and my vision of her is standing outside on a sunny day with a smile on her face, a bushel of clothes at her feet, a cotton clothesline above her head, and a few, worn, wooden clothespins in her hand. Wherever I've photographed clothes hanging in the open, I've been brought back to those early moments in the Pittsburgh yard.

Aside from my lifelong attachment to clothesline images (and the stories they evoke), the time seems to be right for images depicting the pleasures as well as the political correctness of hanging out our wash. I hope my photographs release a few of your memories to some time or place where laundry touched your heart, where you experienced gratitude for Mother Earth as she warmed your soul and dried your clothes. ○

After the Dive

Tree Lines

Yalle Eku Ave Laundry

San Miguel, Mexico

Rose Robinson

WALKING ON THE LIFEBLOOD

Oh, it is so easy to believe
that we are not reliant on dirt,
that we rule the things we stand over
and walk upon.
One forgets that the world is round
and life is circular. And that
a mother survived to give us life
thanks to the dirt that grew her food.
And that we survive
thanks to the dirt that grows our food.
Yes, it all seems miles away
from the seats in our car, and so
in our concrete and pavement lives
it is so simple to forget
that we will all be dirt one day.
But thank god
that the world below our own
is not hell, but dirt,
teeming with cities.
I will not kiss the ground at your feet,
but the soil in which
my grandfather planted trees.

Frances Benjamin Johnston *George Washington Carver, full-length portrait,*
standing in field, probably at Tuskegee, holding piece of soil   159

BULLETIN NO. 31   **SEPTEMBER, 1921**

By G. W. CARVER, M. S. Agr., Director

How to Grow the Peanut and 105 Ways of Preparing It for Human Consumption

Of all the money crops grown by Macon County farmers, perhaps there are none more promising than the peanut in its several varieties and their almost limitless possibilities.

Of the many good things in their favor, the following stand out as most prominent:

1. Like all other members of the pod-bearing family, they enrich the soil.

2. They are easily and cheaply grown.

3. For man the nuts possess a wider range of food values than any other legume.

4. The nutritive value of the hay as a stock food compares favorably with that of the cow pea.

5. They are easy to plant, easy to grow and easy to harvest.

6. The great food-and-forage value of the peanut will increase in proportion to the rapidity with which we make it a real study. This will increase consumption, and therefore, must increase production.

7. In this country two crops per year of the Spanish variety can be raised.

8. The peanut exerts a diatetic or a medicinal effect upon the human system that is very desirable.

9. I doubt if there is another foodstuff that can be so universally eaten, in some form, by every individual.

10. Pork fattened from peanuts and hardened off with a little corn just before killing, is almost if not quite equal to the famous Redgravy hams or the world renowned Beech-nut breakfast bacon.

11. The nuts yield a high percentage of oil of superior quality.

12. The clean cake, after the oil has been removed, is very high in muscle-building properties (protein), and the ease with which the meal blends in with flour, meal, etc., makes it of especial value to bakers, confectioners, candy-makers, and ice cream factories.

13. Peanut oil is one of the best known vegetable oils.

14. A pound of peanuts contains a little more of the body-building nutrients than a pound of sirloin steak, while of the heat and energy-producing nutrients it has more than twice as much.

Makshya Tolbert

LATE EULOGY TO MY GRANDMOTHER

An Ode to Peanuts

I am fixated on peanuts and how they used to look in my grandmother's hands. It took four years after she died for me to prepare peanuts the way she did, as a creamy peanut stew with braised oxtail, sweet plantains, and homemade *piment*. Between my grandmother and me, peanuts were ritual. She would tell "family secrets" while watching me eat her soup at our dining room table, making sure to remind me "not to tell my sisters" what she was about to say. It's been five years since she died, and I'm just now getting to her eulogy.

My grandmother treated peanuts like they were ancestral medicine, a salve. She brought peanuts with her when she left Cameroon to build a family in the United States. Peanuts helped her build home in the wake of leaving home. Peanuts were reparative for me, too. They are the food I remember wanting most as an angry kid experiencing more transition than I could handle.

Needless to say, my grandmother went to great lengths to find the best peanuts possible. Even better was when great peanuts were brought to her. Every December in high school, I was sent home with beautifully roasted Virginia peanuts. They were always gifts for my grandmother. A childhood friend brought them to me sent by her mother, and my job

was to "make sure your grandmother receives them." A relative once brought a hefty bag of peanuts on a visit from our village in Banka. I feel lucky I'm still able to remember her response: "the best peanuts in the world." I miss seeing my grandmother smile. She loved gifts that opened her up.

She taught me to love peanuts for their shape, the brown seed coat that felt papery in my mouth. She made peanuts animate. She ate them slowly, one by one, taking deep breaths and accidental naps in between. I'd wake her and she'd start right back up. Other times, she just popped them into her mouth by the handful. Playful moments. She would snack on peanuts while rounding out stories with an idiom I haven't heard elsewhere to this day: "Even in hell, there are angels."

For a stew, she'd fill the blender with peanuts and blend them until they transformed into a smooth butter. Sometimes she spoke over the blender, her voice equal parts commanding and graceful as she spoke through what she was doing. The motor filled the air, the aroma of peanuts everywhere. She added peppers, salt, garlic, and things I can't write here.

Years eating peanuts and I never quite knew their origins or how they found their way to my life. It took my grandmother dying for me to unravel the history of the food that drew us closest. I find peanuts on the plantation and in slave gardens. I am comforted knowing my ancestors foraged them, and that peanuts found their way into my most cherished memories. I am humbled to learn that the enslaved, not slave traders, brought peanuts with them across the Middle Passage. The Middle Passage haunts me. Peanuts give me something to hold onto.

George Washington Carver innovated peanuts as an offering, as a salve for struggling soil devastated by cotton. The gesture was rooted in love, and Carver integrated peanuts into community life in three hundred ways, as food, oil, moisturizer, and medicine.[1] I wonder if peanuts can decompose some of the trauma carried by the land. I wonder if honoring him as the "father of modern agriculture"[2] can generate a renewed sense of belonging for those of us who have lost so much.

When my grandmother died, I asked myself how to survive in a world without her. I decided to eat what I could remember. Peanuts felt visceral, yet I couldn't bring myself to touch them. For four years, I ate other foods from home: oxtail, vegetables, cassava, fried fish. I moved toward peanuts at the speed of grief.

When I did finally sit and shell my first peanuts in years, I did so knowing this soup shouldn't even be called by the same name as hers. I wept. Five peanut harvests have begun and ended since my grandmother died. Today I am laughing at my grandmother's love of peanuts, how she managed to fit more fully shelled peanuts than you would think possible into the pockets of her favorite matching denim two-piece. Her memory, which is home, is everywhere and anywhere peanuts grow. ○

Notes

1. George Washington Carver, "List of Products Made From Peanuts by George Washington Carver," *Carver Peanut Products*, Tuskegee University.
2. August Brown, "Apocalypse Survival Skill #4: Braiding Seeds," *How to Survive the End of the World*, April 23, 2020.

Carmela Wilkins

BLACK FEMALE SERVITUDE THROUGH THE LENS OF AMERICAN FOOD CULTURE

Recipe Cards

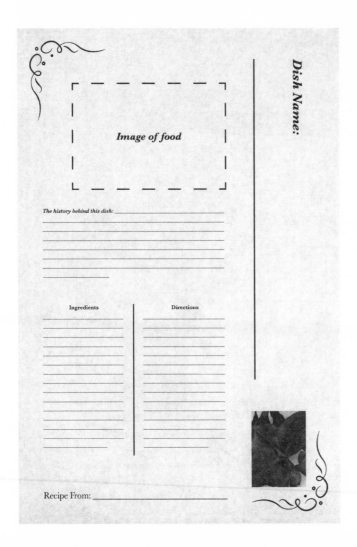

This project aims to archive the narratives of black women historically entrapped in service positions by exploring the relationship between female house slaves and the masters' households where they provided nourishment.

In the consumption of her bodywork, the enslaved black woman was forced to establish a sense of identity, a presence that has embedded itself in modern-day American foodscapes and cookery. The recipe cards are an extension of this relationship, with the dual intention to provide a historical perspective on familiar and unfamiliar dishes and to encourage audiences to archive their own. ○

Okra Soup

*O*kra travelled around the world due to the portuguese slave trade but for most Americans it's a southern vegetable. Okra's origins in the USA are in response to the slave trade, bouncing from the Caribbean to the south. It was heavily used in west african cuisine and still is to this day. All the main points of southern food touch upon okra.

1 qt. Chicken broth
1 qt. beef broth
2 Medium white onion
Large red onion
Fresh cloves
1/2 lb. Tripe (shreds of beef towel)
Hydrogenated oil and butter
1/2 lb. Okra
5-6 garlic cloves
8 small tomatoes
1/2 lb. Smoked Turkey
(All seasonings to taste)
Fresh parsley
Kosher salt
kitchen pepper
Black pepper
Sage
Thyme
Rosemary

Put a few cloves in one white onion. **Stew** chicken broth, beef broth, with white onion and beef towel for 1.5 hours. **Cut** red onion and other white onion in chunks with diced garlic. **Fry** in hydrogenated oil and butter until brown. Cut up okra, **cube** tomatoes, and put into broth. **Add** in the following: kosher salt, kitchen pepper black pepper, sage, thyme, rosemary, smoked turkey, and fresh parsley, **cook** it down for an hour.

Recipe From: _____

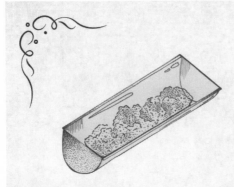

Trough Mush

*T*rough mush illustrates the horrendous conditions that told the enslaved Black person that they were less than human from the days of their earliest childhood until the end of their lives.

Fredrick Douglas quotes, "Our corn meal mush, which was our only regular if not all-sufficing diet, when sufficiently cooled from the cooking, was placed in a large tray or trough. This was set down either on the floor of the kitchen, or out of doors on the ground, and the children were called like so many pigs, and like so many pigs would come, some with oyster-shells, some with pieces of shingles, but none with spoons, and literally devour the mush. He who could eat fastest got most, and he that was strongest got the best place, but few left the trough really satisfied (62)"

4 1/2 cups of water
1 cup of buttermilk
1/2 teaspoon of salt
1 ½ cups of white stone ground cornmeal
Bits of greens, pot liquor, other like tidbits, optional

Let the water and salt **dissolve** together. Add the milk and bring to a **boil**. In small amounts, **add** the cornmeal until it is completely incorporated into the liquid mixture, turn down the heat and **cook** for at least a half an hour. When the mush is near its end, **add** the leftovers. **Eat** with an oyster shell.

Recipe From: _____

Sauce

*T*raditional sauce combination usually paired with meat for dipping, coating etc. West African history has a long history of roasting, grilling and jerking the meat of their cattle, goats, sheep and game meats.

1/2 stick of unsalted butter

1 large yellow or white onion, well chopped

2 cloves of garlic, minced

1 cup of apple cider vinegar

1/2 cup of water

1 tablespoon of kosher salt

1 teaspoon of black pepper

1 pod of long red cayenne pepper or

1 teaspoon of cayenne pepper flakes

1 teaspoon of dried rubbed sage

1 teaspoon of dried basil leaves or 1 tablespoon of minced fresh basil

1/2 teaspoon of crushed coriander seed

1/4 cup of dark brown sugar or 4 tablespoons of molasses (not blackstrap)

Optional: Carolina Mustard Sauce— add 1/2 cup of brown mustard or more to taste, and a bit more sugar.

"Red Sauce"—add two cans of tomato paste or four very ripe red or purple heirloom tomatoes (Large Red, Cherokee Purple, Brandywine, Amish Paste, cooked down for several hours on low heat into a comparable consistency; and two tablespoons of Worcestershire sauce)

Melt butter in a large saucepan, add onion and garlic and saute on a medium heat until translucent. Turn heat down slightly and add vinegar, water, the optional ingredients, and the salt and spices. Allow to cook gently for about thirty minutes to an hour. To be used as a light mop sauce or glaze during the last 15-30 minutes over the pit of coals and as a dip for cooked meat.

Recipe From: _____

Fried Chicken

*C*olonialism revolved around policing every part of enslaved people's lives, including accessibility to eat certain animals, giving birth to slaves' mastery over chicken. Whites had control over cows, pigs and other livestock. The landscape of wild fowl in America during colonialism was not the commercialized commodity of today, instead it chickens were only taken care of and owned by women and children. They were undervalued by society and left to roam freely which resulted in mass breeding, causing greater discontent towards them — the 1800s saw an estimate of approximately 100 million chickens in the United States.

2 White or yellow onions, roughly chopped

4 Tablespoons of lard, shortening, or cooking oil

1 Teaspoon of salt

1 Teaspoon of red pepper flakes or one crumbled pod of long red cayenne pepper

Dried herbs to taste (sage, bits of rosemary, thyme)

1/4 cup of bits of country ham or smoked turkey (optional)

1 medium pan of yellow cornbread (cool, slightly dry, and crumbled)

1/4 cup of chicken, ham, beef, or vegetable stock (you may choose to use the pot liquor from greens as your stock)

Cut the chicken up, separating every joint, and wash clean. Salt and pepper it, and **roll** into flour well. Have your fat very hot, and **drop** the pieces into it, and let them cook brown. The chicken is **done** when the fork passes easily into it. After the chicken is all cooked, leave a little of the hot fat in the skillet; then take a tablespoonful of dry flour and brown it in the fat, **stirring** it around, then pour water in and stir till the gravy is as thin as soup.

Recipe From: _____

Nikki Lastreto

SACRED SPACE, SACRED LAND

Ganesh at the gate. Buddha on the middle path. Archangel Uriel in the Enchanted Forest. Dancing Shiva in the clearing.

The land itself is blessed. With these statues, we amplify the hallowed sanctity of the ranch in the far north of California where we live and grow and flourish. They are reminders of the sacredness of all things.

For many years, I designed sacred spaces for homes and events. My primary approach was to put many deities together—Christian, Hindu, Buddhist, Muslim, Egyptian, Pagan, and more—to teach the simple lesson that if the gods can get along so well, so shall we humans. This is a lesson we all can stand to learn a little more deeply.

Prayer flags hang between ancient Douglas fir and oak trees. A peace sign made of fallen limbs stands in the mixed forest. Our cannabis garden is planted in the sacred geometrical shape from India known as the Sri Yantra. The old growth oaks and Doug fir tower high into the heavens, creating their own inspirational cathedral—a forest.

Sacred spaces are scattered all around the 190-acre ranch my husband and I call home. They attract seekers of all types. Marriages and rituals are performed here, healings have taken place in the powerful stone labyrinth, and celebratory gatherings making music and magic are a natural addition.

There is no doubt that Native peoples also danced on this land in the heart of the Emerald Triangle, and performed their ceremonies under the tall trees and in the big meadow. Their presence can still be felt under our feet and in our hearts. We have been told by visiting elders from the local tribe, known as the Wailaki, that because of the big open spaces here on this hill, it certainly was a sacred space for their ancestors. They say they would have come to gather the acorns from the oaks every autumn, and to then soak them in the bordering creeks. It also would have been the time for families to mingle, marriages to be made, and prayerful dancing to have graced the earth.

It is our duty to carry on these traditions, in our own ways. To that end, we also dance on our land and invite our friends and family to share music and wisdom here. We have had Hindu ceremonies and Buddhist meditations, with Christ and Bastat looking on. Ganja Ma, the goddess of cannabis, lives in our garden and blesses our plants every season as they flourish under the full sun, moon, and stars. She is the patron deity of our crop and never lets us down.

I was raised a devout Catholic; as a child, I went to Mass at least four days of the week. For Christmas, I asked for statues of the saints

← **Nikki Lastreto** *Big Buddha*

Chris Tucker *Sri Mukambika Temple*

all table-top versions. (By then, I was shipping one-ton granite sculptures back to California.) Clearly, I needed a sanctuary.

So, I guess you could say that it was the statues that brought me to living here on this sacred land. I am so grateful. In the summer, I dress them up with flowers and finery. During the winter months, they are simple and stunning in the snow. They are always there, stoically guarding the space and welcoming visitors at the same time. The ancient trees may not speak out loud, but I sense that they approve, as do the creeks, the lake, and the land itself. ○

rather than Barbie dolls or teddy bears. During Lent, I would close the curtains in my room and spend hours on my knees before my altar, praying for a vision. When the Second Vatican Council happened in the 1960s, and the Mass began to be recited in English and the main altar was turned around (and so many more modern changes affected the Church), I was dismayed. How I missed the Latin hymns and the mystery they imparted, the incense and the magic carried down from centuries before.

It was several years later, when I made my first trip to India and entered a Hindu temple, that it all came back to me. Bells clanged, exotic smells of incense and burning ghee delighted me, and ardent devotees chanted in languages that were very foreign. It was intoxicating. I felt right at home, and so began my "icon addiction" once again. While I continued to collect Christian effigies, statues from other traditions were added to my treasures. After several years traveling the world, I had amassed a large collection of icons—and they were not

Nikki Lastreto *Ganesh at the Gate*

Ben Prostine

EQUINOX

The work day is done. The ridgetop
trees are in a fog. A fallen leaf splits
against the windshield and what was it
all for? There were boxes of vegetables
stacked high on pallets, destined
for the warehouses of the untaxed.
There were broken carrot roots, soggy bell
peppers, spilt onions rolling across the floor
and a round sky ruled by the scales.
But the equinox means nothing to the scales
of the pack house, tho we shout the name
of the sign thru the blare of a root washer:
veinticinco libras en la caja, dos libras en
la bolsa, cuántas libras? cuántas más?
The soil runs muddy down the drain,
the fog drops with night and from the top
of the ridge, it's all downhill from here.
Go home, cook a quick meal in the dark,

clean up, shut the coop, read a few pages
before falling asleep and in your dreams
do it all over again: the twelve-hour work day
and the dream shift at night. And where
do all those vegetables of our dreams go?
What tables? What plates? grinded by which
trash disposal? There are no paystubs to give
these hours account. Say it: the scales of
the world are broken. Measure it in *libras*,
court rooms, hunger, debt, dispossession, —
the weight of a six-hundred-foot potato
bed on your back. A scorpion hides in
the hard frosts of November, the interest
on a farm loan grows all winter, and the best
advice when shoveling compost onto the belt
of a spreader is *Don't Stop*. These are
the days when light and dark are equal
but this is a time when little else is.
An hour before dawn the alarm clock rings.
You awake—hands sore from folding the ten
thousand boxes that were never delivered.

→ **Paola de la Calle** *Monsanto mata maiz*

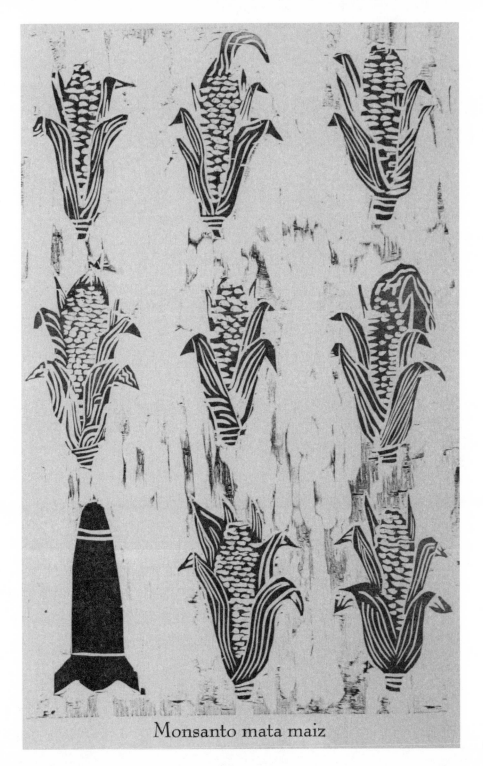

Monsanto mata maiz

Elizabeth Hoover

THE INDIGENOUS CORN KEEPERS CONFERENCE OF UCHBEN KAH

"Mother corn is asking us to rescue her, so that's what we're doing," Omar Raquena proclaimed.

An educator and farmer from Toledo, Belize, Omar was a co-organizer of the Uchben Kah Indigenous Corn Conference that took place in January 2020 at the San Francisco de Jeronimo Roman Catholic School of Pueblo Viejo. I was fortunate to join a group of eight Native

American farmers at the conference, which was attended by over ninety Indigenous farmers from across Belize and Guatemala. Clayton Brascoupe (Mohawk), founder of the Traditional Native American Farmers Association (TNAFA) and the Indigenous Sustainable Community Design Course, collaborated with Omar to coordinate the corn conference that brought

→ *Clayton Brascoupe and Omar Raquena worked together to organize the corn conference*

together a group of representative Indigenous farmers to share their experiences and create a declaration that represented their concerns and aspirations for their profession and their communities. As Omar explained, "We are planting a seed; we need to water it, need to give it adequate elements to grow. For our future, our kids, our grandchildren."

For three days, the assembled group of Indigenous farmers shared cultural information (mostly in English, some in Spanish or Maya with translation) about corn. They described the challenges they face preserving native varieties of corn, discussed their concerns about hybrid and GMO seeds as well the chemicals that many farmers apply to their fields, and brainstormed ways to support each other in maintaining heirloom varieties and increasing youth participation in planting.

The conference offered the opportunity for cultural exchange between Indigenous communities from very different places, all with corn as a key feature of their culture. Sixty-two-year-old Sylvester Tesecum, who spoke in Maya with a little English sprinkled in, has been growing traditional corn since he was eighteen. With another participant translating, he told the story of the origins of this corn that he holds dear.

One day, a woodpecker peered down and saw a group of wee-wee (leafcutter) ants marching one by one, carrying white kernels of corn to stash under a big rock. The woodpecker knocked his bill against the rock, but he gave up—it was too hard for his bill. He decided he was going to need help if he was going to eat like these ants. He went to the *chaak*, the god, and asked him for a favor. The chaak rumbled, "Tell me, what's your problem?" There was thunder and lightning. The woodpecker asked him to break the rock. A flash came— *bam!*—and it split to pieces. The corn that was underneath did not get burned and remained as white as the cob that Mr. Tesecum held up at the conference. The ones that were only slightly burned became reddish yellow; the one that burned substantially became black corn. The

This crop came to these different Indigenous nations in myriad ways and was seen by all present as critically important to their identity and health. Eugenio Ah from the village of San Antonio, Toledo, stated, "When I'm working there [in the fields] and I come up to the house, I drink corn drink to refresh, I feel all the fibers of my body being charged." He is not satiated in the same way by other foods. And it is eating these heirloom varieties of corn that is seen as essential: Terrylynn reflected that "if we don't watch what we eat, we will no longer be Haudenosaunee Mohawk people. We will no longer have the DNA of our ancestors. What you eat over time changes who you are." Peter Raquena, from San Pedro Columbia, Belize, was vehement when he said, "We have to go with our local corn, creole corn. *Maíz negro*. It has lots of protein, something good for me to preserve."

Mario, from Rabinal, Baja Verapaz, Guatemala, represents the Achi Maya Asociación Qachuu Aloom "Madre Tierra," a community association that is working to preserve their original seeds. Speaking in Spanish with Omar translating into English, he described how their original blue corn is healthier for their bodies and not as susceptible to the weevil as the hybrid corns. While farmers in his community get more money per acre to sell hybrid corn, Mario said people in his region prefer the traditional corn for home consumption.

While GMOs are not currently permitted in Belize, farmers from North America gave warning of their impending arrival, and what follows. Terrylynn recounted having to turn under a field of blue corn because her neighbor planted GMO corn and she was concerned about cross-contamination. She advised the crowd, "Fight till you die to keep GMO away from corn. You don't know what you're losing, that's the ancestor of life." A former teacher,

woodpecker was close by when the rock was blasted, and that's why he has a red head.

The Mohawk people, represented by Terrylynn Brant from Six Nations reserve in Ontario, Canada, believe corn grew from the body of Sky Woman's daughter after she fell from the Sky World to begin life on Turtle Island. Eliseo Curley, a corn grower, sheepherder, and weaver from the Navajo Nation, told the story of how Turkey saved everyone from a flood brought on by Coyote stealing the water monster's baby, and then provided all the colors of corn by shaking them from his feathers.

Because of these sacred origins, these original corns have ceremonial value. Felicita Cantun, a Maya farmer from northern Belize explained, "Corn is special to us Mayas—we use it as a food, we use it for ceremonies. We use the four colors during ceremonies—red for east, white for north, black for west, yellow for the south. Four colors. Also, the corn is not just food but also medicine."

Terrylynn also suggested that teachers create a project where students map and document what kinds of crops and seeds families are currently planting. By creating a database of sorts, the community will have documentation if multinational corporations and GMO crops disrupt its food systems.

Janene Yazzie (Navajo) presented information about the International Indian Treaty Council and her and the council's work at the United Nations to advocate for Indigenous rights, and the implementation of the United Nations Declaration on the Rights of Indigenous Peoples. She explained that GMOs don't just create ownership of seeds—they change the memory of seeds. "These seeds don't know where they come from," she said. "They are of no use to our bodies." Peter Raquena agreed with the gravity of the impending threat: "The GMO is coming around and it's going to destroy all of our ancestor seeds."

Additionally, concerns about chemical fertilizers and herbicides contaminating the soil and environment were voiced. An interesting discussion ensued about time—resisting the urge to use chemicals or hybrid seeds because they grow faster or eliminate the need for hand weeding. The tradeoff is the loss of more nutritious and culturally important foods. As Eugenio remarked, "Science really is changing the way we are living when it comes to food production." He reflected on how food like tomatoes that are grown with chemical fertilizers don't taste as good as organic tomatoes whose plants feed off of biomass.

The farmers also shared planting techniques. Felicita described how, rather than planting corn in rows, they "plant in spiral, going like a *caracol*." For the other Maya farmers, much of the planting advice centered around how to avoid cross-pollination between traditional and hybrid corns. Mario from Guatemala saves his seeds from the center of the cornfield, and eats the outer rows which are more likely to be contaminated. Peter advised his fellow farmers that if they plant their corn at least a month apart from when their neighbors plant hybrid corn, it will be less likely to cross-pollinate. Terrylynn from Six Nations plants sunflowers around her field to try to catch some of the GMO pollen that might migrate in from neighbor's fields.

For the farmers in the American Southwest, one of the biggest challenges is lack of water—a situation that starkly contrasted with the pouring rain outside the conference for three full days. Brian Monongye explained how Hopi farmers rely on snow in winter, so when spring comes around they have moist soil to put seeds in. They sometimes plant seeds as deep as twelve inches in the ground, the root system then reaching down until it finds moisture.

↖ *Tamales and tortillas made from local corn*
→ *Lunch*

Field trip
The conference concluded with a series of field trips, starting with Mr. Raphael's field, planted all by hand. He's growing yellow corn and three types of beans: kidney beans, black beans, and small red beans. Some of the land was cleared using slash and burn and then slash and mulch for a few seasons. Every seven to eight years he will rotate to a new patch of land to let this land rest.

When they plant the seed, they don't irrigate because they want the corn's root to grow downward, to become resilient and strong. Eliseo Curley (Navajo) described farming in the Shiprock, New Mexico area using the river to irrigate. Robyn Pailzote from the White Mountain Apache tribe explained how her family does dryland farming, relying mostly on rain. In some places they bury clay pots, which they fill with water that then slowly seeps out during droughts. They pull weeds out by hand or mix black walnut with water to deter weeds. Corn is one of the few plants that can coexist with black walnut.

One of the main foci of the conference centered on the importance of preserving the integrity of local seeds and establishing seed sovereignty for Indigenous communities. As Eugenio stated, "My goal is food security, but I want to add that for me I want to see seed sovereignty [more] than food security. . . . So we can control our source of food." This would entail seeking technical support to create local seed banks and educating farmers about how to prevent cross-pollination of native varieties with hybrid varieties. As Mr. Tush, representing Belize's Agriculture Department, stated, "We need to educate farmers that local and native seeds are the best. We need to bring back seed sovereignty—we need to embed that into us again." Clayton Brascoupe, one of the conference organizers, described the work in his home state of New Mexico to create GMO-free zones, as well as the importance of having seeds stored in multiple places in case one site is destroyed. He told the story of a farmer whose seed shed was destroyed in a fire, but who was able to get all of his seeds back because he had shared them with his neighbors over the years. Part of building seed sovereignty for a community entails farmers connecting and sharing in this way. According to Terrylynn, it also entails farmers taking control of the breeding process again. "I do not believe we as Indigenous people have lost anything. People say our seeds are lost. I think we've fallen down in our responsibilities as seed keepers. I have four varieties I'm breeding right now. If every farmer took the gumption to breed, we'd be back to hundreds. That's our obligation as farmers—to keep that going," she said.

One challenge that the Maya farmers described in maintaining traditional farming practices is that their young people go to schools where they are taught in English, and are mocked for speaking their Indigenous languages. Increased reliance on the wage economy also means youth aren't often willing to help in the fields unless they are paid. But while many expressed the challenges of encouraging youth to take up farming, sharing Indigenous knowledge with future generations was seen as paramount. This would entail making seed saving an intergenerational activity, and could involve local schools developing curricula that include traditional corn.

Eugenio is working to train his ten-year-old grandson to be a farmer, bringing him along to all of the workshops and trainings he attends, including this conference. He relayed a heartbreaking story about attending a career day in

Punta Gorda where none of the students were interested in hearing about farming—they directed all of their attention towards the firemen, police, and doctors. "I say anybody hungry? I tell them who is going to feed the policeman, nurses. Everybody went silent. After my presentation I have two small kids beside me. Hurts my heart, I wanted to cry." He encouraged the parents and educators in the crowd, saying, "We need to look at farming as a career, a calling."

And as Brian pointed out, part of inspiring the youth is connecting them to the ancestors. "Let's continue to plant that seed for our future generation but also keep walking into the prayers of our ancestors that have put those prayers out for us to keep going forward. We have to fall back on our histories of relationships as people. Protect our land, territory, and resources. Perpetuate this knowledge that was given to us in the beginning," he said.

In spite of the importance of their work, many of the farmers expressed that they do not feel respected or valued in their profession. As such, they need to gain government recognition as a group whose voice should be recognized. Along these lines, a plan was developed to form committees or an association of farmers to represent their communities and concerns, draft resolutions to protect their seeds, present a united front to better resist multinational organizations like Monsanto (Bayer), and host more conferences and gatherings for farmers to learn from each other. Gatherings like this corn conference were identified as important opportunities to network, to share ideas and solutions, and to bring information home to their communities. At the conclusion of the second day, a group was formed and tasked with planning additional events and community statements about the preservation of heirloom seeds.

The final day of the conference was dedicated to workshopping the Declaration of Indigenous Corn Keepers Conference (Uchben Kah). Janene created the first draft based on two days' worth of notes, and the group spent the morning reworking it to reflect what they felt were the most pressing issues, and what would be best received among their colleagues. The first agreed-upon draft is included below, but should be considered as a living document.

Colleen Cooley sharing corn seeds

Declaration of Indigenous Corn Keepers Conference (Uchben Kah)
Starting Draft*
January 19, 2020
Pueblo Viejo, Belize

We, as Indigenous Peoples from Queqchi, Mopan Maya, Yucatec Maya, as well as Navajo, Hopi, Mohawk, and Pueblo communities from North America who have gathered at the first Uchben Kah Corn Conference, come to this sacred territory to share our knowledge and wisdom, and to plan for the future of our communities and our Peoples.

As farmers our lifeway is rooted in growing our own food and protecting the seeds that have been passed down for generations. Our farming teaches us many ethics, like how to organize our communities and governments, and how to be proud of who we are. We are connected here through the recognition that corn is who we are as a people, and a desire to bridge the networks between our communities that have been severed. Together we shared stories about the importance of preserving traditional farming practices and protecting traditional seeds from the threats of hybrid seeds and Genetically Modified Organisms (GMOs).

Even though we come from different regions, from different ecosystems and environments, we acknowledge that we are all connected through our corn. We also acknowledge that we have lost so much of our traditional lifeways, but we will no longer allow this sacred food to be threatened by hybrid or Genetically Modified Organisms promoted by those who do not honor the sacred memory of our corn. We commit to continue to build our power so that we can secure the future integrity and purity of our local seeds as Indigenous farmers.

Towards this end we have developed this Declaration to outline our shared commitments moving forward and to exercise our inherent rights as Indigenous Peoples.

We declare:

1. We will commit to continuing the traditional and cultural practices of our ancestors that connect us as people, in the protection and propagation of our sacred corn seeds, and all our traditional foods and medicines that our people have depended on for generations. We know that to feed ourselves requires us to center our responsibilities to our families and our communities in the growing of our food, which includes revitalizing the traditional teachings of our ancestors and passing them on to our children. We do this to honor that the knowledge we have from our ancestors is a gift from our creator and a gift from our mother, the corn.

2. We commit to connecting elders with young people through our traditional language, which is the essence of who we are as people.

3. We will integrate these teachings and practices into the educational systems from the primary school level so that our youth will know more about traditional farming.

4. We will form an open committee or association to represent a bigger body that can represent the concerns of all our farmers.

5. We will locate keepers of local traditional seed varieties and propagate them so to be able to share and protect our seeds with farmers.

6. We will encourage farmers in our communities to utilize native local seeds, emphasizing seed sovereignty and traditional farming practices.

7. We will acquire technical and financial assistance in the development of local seed banks in order to make traditional seeds more available to our Indigenous communities.

8. We will seek recognition from local and federal governments of the issues that are important to Indigenous farmers.

9. We will elevate the status of farmers because food is important—not just commodity farmers but true farmers that raise animals and other foods and medicines. ○

*This statement was developed at the Corn Keepers Conference on January 19, 2020, but should be considered a living draft, as participants of the conference wanted to take this document home and get feedback from their fellow community members about its content. The intention is also for this document to be translated into Spanish, as well as the Indigenous languages of each participant's home communities.

Photos by **Elizabeth Hoover**

↑ **William Henry Fox Talbot**   *Dandelion Seeds*
→ **Edward S. Curtis**   *Gathering Seeds–Coast Pomo*

Seed Range

USDA, Botanic Garden Land

Tony N. VanWinkle

WEEDS AND/AS ANCESTORS

Coevolution, Memory, and the Ambivalence of Garden Maintenance

It happens every year. The garden begins as usual. We clear winter debris—last year's ghosts—and work the soil, preparing beds. The point is to start the season anew with a clean planting surface.

Then come the weeds, always growing faster and with greater profusion than whatever was intentionally planted nearby. Even as we bemoan their emergence, many of these plants might be understood not as trespassers, but as companions. These anthropogenic plants are annuals that thrive specifically in human-disturbed habitats. As such, they have a long history of coevolution with humans.

"Weeds" remind us that our symbionts continue to pay attention to our kinship even when we do not. In Vermont's Northeast Kingdom where I live—US Department of Agriculture hardiness zone 4a—these include chickweed, lamb's quarters, various wild amaranths, and purslane. Most have traditional culinary and medicinal uses: Much-maligned galinsoga (also known as quickweed), considered noxious even among many organic growers, is one of the main ingredients in the Colombian variation of the one-pot meal known as *ajiaco*. In a traditional Mesoamerican agricultural system, these food-producing plants are known generally by the name *quelites*, which refers to all un-

cultivated edible greens. Although built around the intentional planting of maize, beans, squash, and chiles, this intercropping system, commonly known as *milpa*, also encourages anthropogenic weeds, including as many as 127 species of quelites.[1] However, these plant foods have become increasingly underutilized as discourses and practices of dietary modernization have resulted in the stigmatization of undomesticated "weeds" as food sources.[2]

Just before her untimely death, my wild-crafting mentor authored a self-published book, *Eat Your Weeds*. In addition to regular foraging forays in the ditches and margins of red dirt roads in the Oklahoma tall grass, we

Perennial Goosefoot *(Chenopodium Bonus-Henricus)*

shared garden space for a time in the Franciscan community on the prairie where we both lived as sojourners. She lived alone, seeking respite from the stresses of living in an old stone house whose fragility was newly exposed by fracking-induced earthquakes. I lived with my family and traded land stewardship and cooking duties for rent while working as a field researcher, studying the ways that local land and life were adapting to the demands of Middle America's "rural transition" from productive agricultural landscapes to bedroom communities composed of ten-acre "ranchettes."

Working side by side in the garden, she regularly chastised me for pulling lamb's quarters out of the ground. "It's better for you than anything else you're growing," she exclaimed. The genus to which lamb's quarters belong is the same one to which the currently trending (though long-cultivated) "superfood" quinoa belongs. Both are *Chenopodiums*, popularly known as goosefoots. Both have a long history of human dietary use, as a source of both grain and greens.

In her poem "The Change," Cherokee poet Allison Adele Hedge Coke[3] recreates a shifting world of rapidly mechanizing tobacco farms in North Carolina through the eyes of "traditional agriculture people and sharecroppers" functioning as de-skilled day laborers. It narrates the "slow murder of method from a hundred years before," a world of dirty, hard work under an unrelenting sun, punctuated by the indignities of chemical exposure and constant reminders of the desacralization of life shared in relation to the land, to other species, and one another. In doing so, the work recalls the world before:

Before edgers and herbicides took
What they   call weeds
. . . .
When we still remembered
. . . .
That then the tobacco was sacred to all of us
and we
Prayed whenever we smoked

Indeed, the clean, monocultural approach that insists on eliminating those species and relationships that "do not belong," manifests not just on Hedge Coke's tobacco farms or in the fields of Corn Belt, but in ways more mundane and insidious: straight lines, clean ground, with no room for anything that isn't intentionally planted. Or no tolerance, perhaps, because the physical expression of the monocultural paradigm duplicates itself internally as well, yielding what Vandana Shiva calls "monocultures of the mind."[4] In such a colonized mindscape, knowledge is among the principal

targets for extirpation. As the knowledge of
the colonizer becomes standardized, incon-
testable—normal, correct—the knowledge of
the colonized is correspondingly cannibalized,
delegitimized. It is through this process that
our weedy companions are transformed from
welcomed food to unwelcomed pest.

Amid such an imaginary, the work of resilience and networks of reciprocity that enable
that work might be increasingly shaped by
the wisdom of weeds, which is the wisdom of
ancestors. As Enrique Salmon writes, "Food
landscapes remain intact when old recipes
are regenerated."[5] And so we encourage, even
nurture the weeds, so that we may recover what
they have never forgotten. ○

Notes

1. Edelmira Linares Mazari and Robert Bye
Boettler, "Las especies subutilizades de
la milpa," *Artículos* 16, no. 5 (May 2015).
2. Mazari, Boettler.
3. Allison Adelle Hedge Coke, "The
Change," *Dog Road Woman* (Minneapolis,
Minnesota: Coffee House Press, 1997).
4. Vandana Shiva, *Monocultures of the
Mind: Perspectives on Biodiversity and
Biotechnology* (London and New York:
Zed Books, 1993).
5. Enrique Salmon, *Eating the Landscape:
American Indian Stories of Food,
Identity, and Resilience* (Tucson, Arizona:
University of Arizona Press, 2012).

Addison Bowe

GRASS GONE TO SEED

The sun was keeping us company today.
We laid in the grass,
sipped nettle tea and looked
out at the sprawling farmland.
Open space can be a great friend.

Addison Bowe

Geoff Nosach

ON WOOD-MEATS
AND LOCAL SENSIBILITY

Wood-meat can help fulfill the growing demands of folks who aspire to eat a more sensible diet. Industrial animal meat production relies upon disproportionate amounts of grazing and cropland per pound, putting undue pressure on the natural balance of species diversity. Grazing land is often the end result and pretext for clearing old growth. Every temperate region of the world is experiencing temporary or permanent deforestation, leading to the extinction of species and degrading the habitat of which we are a part. Forest-born meat, whether animal, vegetable, or fungi, can support stable local food systems and help to end the incremental destruction of our greatest natural resource, a diverse environment. With emphasis put on safe and sustainable methods, we can procure substantial amounts of protein from a forested habitat.

Wood-meats, then, are mostly fruits, nuts, and mushrooms. Fruits are commonly much higher in sugar and fiber than typical meats but some, like avocados, are decidedly fatty and meaty. Nuts and seeds are high in protein and are a great replacement for meat and dairy. Mushrooms are truly the meatiest of the bunch and by themselves can claim the title of wood-meat. They can fully embrace and replace the texture and flavor of animal meats. Chicken of the woods mushrooms, when breaded and fried, are easily mistaken for the real deal.

Many mushroom species can be farmed on trees and raised in multi-purpose wood-meat forests. Oak and beech forests, for example, produce quantities of edible seeds while being an exceptional species on which to grow gourmet mushrooms. The forests of eastern North America are, or were, dominated by these and other meat-producing species like walnut, butternut, hickory, and chestnut. All these also happen to associate with a bounty of native gourmet mycorrhizal fungi, like chanterelles and boletes.

Agroforestry practices can be implemented to produce young growth logs, a ready-made mushroom substrate, and old growth trees for limbing and seed production while maintaining forest networks and mycorrhizal associations. Maple sugaring is an example of a forest-farming practice that is common in my bioregion. Most gourmet mushrooms grow well on sugar maple, and mushroom cultivation may help diversify income on the sugar farm.

Squirrels are a great example of an abundant wood-meat source, happily fattening up all year on tiny, difficult-to-hull seeds and nuts. Deer thrive on the forest's edge and are well-adapted to wooded domestic landscapes.

→ **Poppy Litchfield** *Field Mapping*

The careful regulation of habitats can play a role in wood-meat production, and managers should always consider food systems and species diversity in their valuation. The intersections of hunting and conservation are especially important places for building coalitions between politically and economically divergent groups. Management of hunting, fishing, and gathering has played a critical role in the development of wild habitats throughout the history of humanity. This is where agriculture and hunting-gathering methods intersect; where species nurture from within an ecosystem, rather than dominating from above.

When they are not encroaching on wild populations, many domestic critters can fatten up on forest bounty. Reforested farms could benefit from a diverse livestock rotation. Pigs can forage on roots, fruits, and nuts; ducks, on bugs and slugs (common pests in the mushroom yard); goats, on saplings, twigs, and leaves. Pollarding and coppicing are forest pruning strategies which produce densely spaced hardwood shoots to be harvested for mushroom logs, firewood, and livestock feed. Grazing and foraging herds may put on less fat and be less productive than conventionally raised animals, but their husbandry would require much less imported feed per pound. Less yield per acre would mean more diversified acres to fill the need. Luckily, every acre of wood-meat production can support biodiversity, natural habitats, and ecosystem restoration. Envision the harvest: acorn pancakes smothered in tree nut butter and maple syrup, salty fried shiitakes, and the occasional duck egg or game stew.

Less appetizing is the animal-meat industrial complex. Monocrop grain production spreads across once-fertile plains while feeding is heavily concentrated in environmental wastelands. Lab-made meat which employs the same agricultural methods is not a solution to diversity loss. Clever marketing that claims to heal the world with burgers is "impossible" and should be considered "beyond" rationality. A more sensible remedy is to transition the animal industries toward pastures and forested farms. Eating far fewer animals is a good start.

Let's consume less of everything, decrease and decentralize production. Doing less, we can restore the feedlots to forests, plains, swamps, and savannas. Small flocks and herds can graze naturally, foraging while fertilizing fruit and nut trees. Many impoverished farms still operate in this timeless manner, as a complement to diverse environments. These farms go unrewarded under current economic realities, and a just transition for producers and consumers is anything but certain. Adoption of localized agriculture and foraging networks benefits smaller producers and diversified food systems.

For the privileged modern omnivore, this transition certainly could only come with compromise. Convenient, fast, and exotic food would give way to regionally adaptive staple crops, seasonally available produce, and local wood-meats. The benefits include healthy food for consumers, healthy work for farmers, and bottom-up wealth distribution through small, local economies. Sensible producers are already decoupling agricultural development from the infinite-growth paradigm by pioneering sustainable, organic, and agroforestry practices. The will to reject industrial food products and include wood-meats as a greater portion is prerequisite. It is the will for an intentionally diverse, mostly forested, and meaty landscape. ○

Elicia Whittlesey and Reid Whittlesey

PREPARING AND REPAIRING: A FAMILY FIRE STORY

On November 8, 2018, we received a text message from our dad alerting us to a fire that had started early that morning, ten miles east as the crow flies from his home and ten-acre property. Ever since we moved away, our dad has faithfully sent us updates from the northern California canyon where we grew up. These notes light up visceral memories of bumblebees, mushrooms, wildflowers, and wildfires.

Fires are part of life in a seasonally dry ecosystem like California's. A grass fire burned through the canyon in the early 1990s on Father's Day, searing its smoke, ash, and altered landscape into our early memories. Many other fires have come close, spurring adrenaline-inspired brush clearing and weed whacking beyond standard precaution. They mostly come during the heat of summer and behave roughly in ways we understand: moving slowly, burn-

ing with variable intensity, rarely scorching houses. They leave the land changed, but not beyond recognition. This fire was remarkable for having started in November, which is usually the beginning of California's rainy season.

After receiving the November 8 text, Reid called our dad. He had just filled up jerry cans of gas for the five-horsepower Honda pump that runs our fire protection system, and reported that a plume was growing quickly across the Feather River near Pulga, a region we usually disregard as a potential source of fire for the canyon. Reid went back to work cutting and burning Russian olive out on a ranch in central New Mexico, not anticipating that the fire would develop into a serious threat.

Later in the morning, our dad sent a photo of an enormous smoke plume and the grave news that the fire was in Paradise, three miles upstream from his house. A worst-case scenario was underway, as the fire burned through the small town composed of retired folks and huge Ponderosa pines, with few evacuation routes, sustained high winds, and extremely dry conditions. Everyone felt that a catastrophic fire in Paradise was inevitable, and nearby fires that ignited were usually rapidly suppressed. With the capacity to move across the deep canyon of the North Fork Feather River in about an hour, what kind of fire was this?

Jan Arendtsz *After the Fire*

We texted back and forth, trying to understand the fire's severity and trading updates from our dad. We knew his internet was down, and it was hard for any of us to get information about where the fire was, or how fast or in what directions it was moving. From Colorado, Elicia scanned the internet, and on Twitter she found stories that have come to characterize the Camp Fire: people evacuating the town of Paradise, stopped in traffic, getting out of their cars to run from the flames. Knowing that our dad planned to stay and defend the house and property, we called periodically to convey the impressions of the fire that we could glean from the internet and Reid's contacts in the California fire community. He picked up when the cordless phone was in range, between raking leaves, moving hose lines, keeping the pump fueled, and cutting firebreak. I think this is a different kind of fire, we each told him. The concussion of exploding propane tanks in Paradise had been audible all day, impressing upon

him the gravity of the blaze and the unprecedented damage it was incurring.

An hour after sunset, the fire was within a
few hundred yards of our house. Our dad heard
our uncle's barn a quarter mile up canyon go
up in crackling, popping flames. He made the
judgment call to leave, and at the mouth of
the canyon, he called us on his way to Chico.

"I think it's 50-50 on the house."

For nearly a day, we didn't know if the
house had burned. Our family has practiced
active management on the property for years—
thinning, encouraging fire-adapted native
species, and burning in small, managed fires—
that ultimately kept the damage caused by
the Camp Fire minimal. But these efforts were
not enough to entirely escape its detrimental
impacts. By the next afternoon, our dad had
returned to the canyon to survey the damage.
The bridge over the creek had partially burned,
as had the spring-fed water tank, but the house,
sheds, and majestic valley oaks on the property were standing. It was clear immediately,
though, that many parts of the canyon had
burned, hard, and would never be the same
gullies, side canyons, and oak meadows that
we'd known.

Reid returned to California a week after
the fire to help with the recovery. It was still
actively present in the landscape: there was
no smoke, but until it rained a thin lens of ash
would rise from the ground with any distur-
bance. Pages of books, incinerated and lifted
by the fire's high winds, were found among the
blackened soil and scarred trees. It didn't smell
the same as a wildland fire; there was something acrid and pungent that suggested the
disaster it was. Even old growth blue oak (*Quercus douglasii,* a California species adapted to
withstand drought and fire) had fallen. Because
humidity was so low, approaching two percent

in the days leading up to the fire, accumulated
duff smoldered in the roots and caused these
beautiful old trees to succumb.

In March, we walked the land together, just
as the surviving oak trees put out new leaves.
For some time, the oak's seasonal dormancy
masked the survival rate. Now, we could see
that entire reaches of the creek had lost their
overstory of old valley oaks and California bays.
We imagined fast-growing alder and willow
growing up in the gaps, or maybe invasives
like Himalayan blackberry, Scotch broom, or
Arundo Donax, a gigantic grass species that's
an ecological game-changer. We speculated on
the nature of the ecological changes, wondering if we could influence the course of this
disrupted ecosystem toward faster recovery or
the regeneration of native species.

By summer 2020, it is clear that although
many trees are dead, they are still having an
effect on the land. Our dad complains that his
best swimming hole is gone, the deep stretch
of creek filled in by fallen dead alders and large
old willow branches. These complexes of large
woody debris make perfect spots for sunning
turtles, create refugia for fish, and slow the
water's flow, allowing sediment to accumulate.
No longer known as the swimming hole, the
now-murky pool has gained the name "turtle
slough." We watch as the natural processes
shift to an initial healing response, revealing
that, in many ways, the land will take care of
itself without human intervention. For that to
happen, our neighbors will have to join our dad
in viewing the downed trees as ecologically
beneficial, not as ugly messes to be removed
by chainsaw or tractor. We can encourage the
land's resilience through continuing to prepare
for fire by burning and thinning, but we also
need to know when to get out of the way. ○

To prevent Crows pulling up Indian Corn

Take three or four old shoes, that are worn out, and fill the toes of them with sulphur, or the roll of brimstone broken small, make a fire with chips, or any small dry wood in or near the middle of your cornfield on a flat rock, or on the bare mould, (a rock being preferable) after planting your cornfield, then lay the toes of the shoes on the fire and let them continue until the leather be burnt through, and the brimstone has taken fire; then after sticking down poles of ten or twelve feet in length at each corner of your field, and inclining them towards the centre, make a string fast to the heel quarters of each shoe, and tie it fast to the top ends of the poles, letting the strings extend half way down, and when swinging, not to interfere with the poles; and no crows will alight on your field that season.

From the *Old Farmer's Almanack*, 1804

Alejo Salcedo *Maíz 1*

Yanna Mohan Muriel

DISPERSED SEEDS FROM PUERTO RICO

"A farmer anywhere is a revolutionary." This is the passionate statement shared by a farmer fifteen years ago at Old San Juan's Organic Market, and has been the leading sentiment behind a growing organic movement that has found strength and support within global networks. In Puerto Rico, farmers consciously cultivate the land as an act of political resistance, and today we are fighting for the integrity of seeds. Many decades of public policies eradicating small-and medium-scale farms—a means to pave the way for heavily subsidized corporate control of seeds—have sent a strong message of governmental priorities.

Farmers' intimate relationship with the earth brings us harvests, but it is the seed that unravels the strength and sovereignty to till the colonized land. In 2009, our capacity to protect our seeds was overpowered. Then-governor Luis Fortuño signed into law a policy that promotes and develops the agrobiotechnological sector in Puerto Rico, claiming that genetic modification produces food at lower costs. It is imperative, the law states, to establish favorable conditions for Puerto Rico's biotechnology industry to become competitive in the global market and increase operations related to the agricultural field.[1] It prevents local authorities from regulating in this area and enables the industry's access to basic resources (water, land, incentives) even as local farmers, myself included, struggle to obtain much-needed supplies to produce food.

Over the past forty years, as small-scale family farmers built a local organic farming movement, foreign corporate interests such as Monsanto, Bayer, Syngenta, Dow, Pioneer Hi-Bred, RiceTec and others, were busy transforming Puerto Rico into the world's greenhouse for GMO seeds.

Since the 1980s, the US federal government has granted the territory 1,872 permits for research and development of GMO seeds.[2] This is 282 more than the leading US state, Hawaii. According to Puerto Rico's Center for Investigative Reporting, "Multinationals do not use these lands to produce food, but to experiment with seeds they send out of the country where the research and development process continues, and then they sell them in the global market."[3]

Indignant about the immense social and environmental costs locally and overseas, the local organic farming movement joined international networks of solidarity, advocating for environmental, food, and seed justice. Active members, including our local organization, Boricuá,[4] identified La Coordinadora Latino-

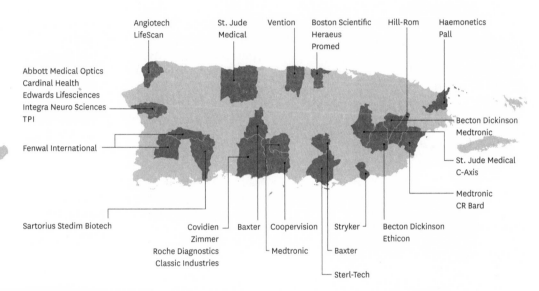

This map, based on data provided by Puerto Rico's Department of Economic Development and Commerce, demonstrates the strong presence of foreign corporations.

americana de Organizaciones del Campo-Via Campesina (CLOC-Via Campesina), the world's largest international peasant movement, as a platform where to voice our struggles.

We needed (and still need) international allies. Once again, our autonomy was being held hostage by the US, so we began to organize around issues of seed sovereignty, native seed conservation, and seed diversity. Since joining these networks, our story has been retold in Argentina, Spain, India, Turkey, Cuba, Perú, Brazil, Italy, North Korea, Philippines, Nigeria, Ghana, South Africa, Germany, Haiti, Indonesia, Palestine, Nicaragua, Colombia, and France, among other countries. Our story has been documented in manifestos that circulate the world thanks to CLOC-Via Campesina, which Boricuá joined in 2013 during their VI International Conference in Jakarta, Indonesia.

In 2014, Boricuá became part of another

international network with a special focus on ecological, biological, biodynamic, agroecological, and organic farming practices: the Intercontinental Network of Organic Farmer's Organizations (INOFO). Since joining these global platforms, local farmers have travelled, participated in advocacy workshops, and exchanged farming practices with other farmers from distant nations and cultures. INOFO, an autonomous structure within the International Federation of Organic Agriculture Movements, was created to address the alarming rate at which small-and medium-scale farmers are disappearing worldwide.

It is not only the farmer who needs to be emancipated from current industries; our seeds are at their most vulnerable. Threatened by the homogeneousness of foods, sparse monocultures, and the distribution of limited varieties, seeds are at risk. Currently, the autonomy and well-being of seeds are being dictated by patents, restrictions, seed exchange bans, and other restrictive schemes developed by a few agro-corporations. "Seed industries, including those marketing GMO seed, lobby

and make policies on an international level, where their interests are pushed through platforms such as the WTO [World Trade Organization], economic blocks, the FAO [Food and Agriculture Organization of the United Nations], and their decisions are implemented locally," Daniel Wanjama said in an interview.[5] Wanjama works with the Seed Savers Network in Kenya, a grassroots network of more than fifty-seven thousand farmers, and is a leading member of INOFO. He believes that it must create a network to advocate for the integrity and well-being of seeds on an international level. The Puerto Rican agroecological movement supports this effort, hoping to offset the responsibility for exporting GMO seeds to the rest of the world. Within these global networks, we hope that one day local governments will abandon corporate ties and respond to the needs of the farmers, allowing us to step away from fighting—or revolutionizing—to fully sow, cultivate, and heal our lands with the vitality of seeds. ○

Notes

1. *Ley Número 62 de 10 de agosto de 2009* (Ley de Promoción y Desarrollo de Empresas de Biotecnología Agrícola de Puerto Rico), LexJuris de Puerto Rico.

2. Eliván Martínez Mercado, "The Boom of Monsanto and other Seed Corporations Blows in the South of Puerto Rico," Centro de Periodismo Investigativo, March 7, 2017.

3. Martínez Mercado.

4. "¿Quiénes somos?," Organización Boricuá de Agricultura Ecológica.

5. Daniel Wanjama, phone interview with author, April 2020.

Jessica Manly

WE ARE EXCITED ABOUT BEANS

Grocery shelves of canned and bagged beans were the first to be wiped clean with the onset of COVID-19 in the United States. When restaurants and farmers markets closed down and stay-at-home orders were put in place in most states, many continued to stockpile beans and other storage staples in an attempt to assuage fears about an unknown future. The US Dry Bean Council reported a 40 to 50 percent increase in demand for dry beans in the first months of the virus.

Though the pandemic has helped cement demand for pulses, heirloom beans in particular were already riding a quiet resurgence among small-scale farmers, chefs, and dedicated home cooks. This was not born of fear of scarcity, but a devotion to their understated flavors, versatility, health benefits, and histories. Napa-based heirloom bean purveyor Rancho Gordo's sales have grown 15 to 20 percent every year since its launch in the early aughts, and they were up 164 percent by June 2020. The company's subscription bean club boasts over eight thousand members and a twelve thousand-person waiting list.

Heirloom beans are open-pollinated, meaning the seeds can be planted and saved year after year, yielding roughly identical genetics. Though they can be challenging to cultivate on a large scale and tend to be lower-producing, heirlooms retain their unique flavor profiles across generations. Despite production challenges, these distinctive traits and agricultural legacy appeal to many young growers.

Caitlin Arnold Stephano first began cultivating heirloom beans on a small farm on Washington's Vashon Island. She got hooked after starting with just a few varieties in 2009. "The farmers I know growing dry beans are doing it because it's just so fun," Arnold Stephano said. "There are so many varieties and they all have interesting stories, like any heirloom crop."

She's partial to beans with names and origin stories as colorful as the seeds themselves: Jacob's Cattle, Painted Pony, Ireland Creek Annie, Olga's Pink, and Kilham Goose bush bean. Kilham Goose was developed by a farmer on Whidbey Island, north of Vashon, and has a shiny, dappled skin of maroon and white, like an Appaloosa pony—also the name of another heirloom bean. "It feels like magic," she said. "You plant one bean and you get so many back. And you can store them pretty much indefinitely."

She's right. Beans, if dried properly, can see you through not just a global pandemic, but millennia. The first wild beans were gathered for food roughly nine thousand years ago, with the first cultivated variants dating back to 2000 BCE. Originating in Central and South America, beans were one of the many crops domesticated by Indigenous peoples and saved season after season. One heirloom variety, the cave bean, was uncovered in the 1980s in a sealed clay pot in New Mexico, and is believed to have been first cultivated fifteen hundred years ago by the ancestral Puebloans who inhabited what is now the Four Corners area. Some of the unearthed ancient cave beans germinated successfully,

Molly Reeder   *Big Heirloom Bean Babies*

and they are widely available from heirloom seed distributors.

Today, the Zuni, Hopi, Iroquois Six Nations, and other Indigenous peoples maintain seed banks to protect the crops essential to traditional ceremonies, diets, and culture. Of the more than four thousand varieties of beans grown in North America, less than 20 percent are available commercially.

Through the Rancho Gordo-Xoxoc Project, founder Steve Sando purchases beans directly from growers in Mexico who cultivate heirloom varieties in an effort to counter the "international trade policies that seem to discourage genetic diversity and local food traditions," he said. The larger volumes Rancho Gordo prepurchases from Mexican farmers allow the farmers some financial security at the start of their growing season and ensure a market for their heirloom crops so that they can avoid only growing conventional bean, corn, and soy varieties for international commodity markets. Sando said creating a market for at-risk crops is the best way to save them. Though the Xoxoc Project has received significant attention from press and consumers, the vast majority of Rancho Gordo beans are actually grown in the US, specifically in California, Washington, Oregon, and soon in Arizona. "Some of these beans are indigenous to this area here in Napa, and most people don't even know what they are. That education piece is always part of the focus— this is a great ingredient that you've taken for granted," he said.

Sando believes the challenge of harvesting often becomes a deal breaker for young and beginning farmers. "We've worked with new farmers who are all gung-ho, and then they just fall apart at harvest time," he said. "We're trying to do a mix of old-timer California farmers, and newer farmers, but the newer ones often

don't quite know what they're in for—we've had several crops we've invested in and the farmers have ended up walking away."

The beans either need to be harvested by hand, which is time consuming or an expensive labor cost, or production must be scaled up to acreage that justifies purchasing or renting costly harvesting equipment. The beans also take up a lot of valuable ventilated storage space for drying in the few months after harvest. Judging harvest time can also be challenging. Last year, Arnold Stephano waited too long to harvest and caught a hard frost, which cut her yield by more than half. Temperature fluctuations and extreme weather events resulting from climate change are only making these decisions tougher and losses more common. She agrees that it's often a tricky proposition for new farmers: "When I tell other farmers I grow dry beans, they're like 'what are you talking about?'"

"It might sound weird that so many people are excited about dry beans—that I'm excited about beans—but you taste them fresh and you just get it," Nick Lubecki, a farmer in Butler, Pennsylvania, said. Nick grows a variety of staple crops with his brother, Justin, including potatoes, onions, corn, rice, wheat, and rye in addition to Good Mother Stallard and other heirloom bean varieties.

Despite his enthusiasm, Lubecki agrees that hand harvesting can be a real barrier to scaling up production. Instead of picking each pod individually, he waits for the beans to dry in the field and then pulls up the entire plant to dry in crates in his attic. He also found an online tutorial that coached him through converting an old wood chipper into a thresher that has minimized his hand work.

Another major challenge for farmers getting into beans, or any other crop for that matter, is access to secure, affordable, quality land—repeatedly the number one challenge reported by beginning farmers and ranchers across the country.[1] Over the last decade, farmland prices have more than doubled.

Arnold Stephano, who currently lives in New York's suburban Hudson Valley, has been getting around the land access challenge for several years through a cooperative arrangement with friends who own a farm in Bethel, Vermont. She splits her harvest with the farm owners in exchange for use of their land and their help with weeding. Even through this arrangement, Arnold Stephano knew she would never be able to turn a profit. "I think if you want to do dry beans lucratively, it would need to be your main focus. You need the storage and equipment. Or you could grow a few other crops that need the same growing conditions," she said. Her dream has always been to start up a garlic, onion, potato, and bean operation, which she jokingly refers to as "Beans 'n Taters Farm."

"To make any money at it, you really have to grow at volume," Sando said. He generally contracts a farmer to grow a single variety each season so they can focus on the needs of the crop and grow at scale. It usually ends up being more profitable than trying to grow several varieties or additional crops.

These production challenges mean that small-scale growers often have to charge super high prices at market to break even. "Some of my friends who sell at farmers markets have to charge about $10 per pound or more," Arnold Stephano said—a tough sell when up against conventional dry beans which can cost as little as two dollars per pound at the grocery store. Rancho Gordo heirloom beans go for six or seven dollars per pound on their online store. But cost doesn't seem to be a factor for customers who pay a premium for the jewel-colored goods, the connection to the farmers who grew them, and the chance to participate in the legacy of seed-saving.

Rick Easton, owner of Bread and Salt in Jersey City, New Jersey, hosted a weekly pop-up called Bean World at Bitter Ends Garden & Luncheonette in Pittsburgh, Pennsylvania. The entire premise was to let the inherent flavor of the legumes speak for themselves, and on Friday evenings, he served cooked beans straight up with olive oil and salt. Lubecki was one of the suppliers for the events and said, "Everyone was given two types of beans to pick from and that was it. I think eventually someone talked him into giving you some bread too."

Arnold Stephano prefers to cook them simply as well, boiled with a few dried chilis, cinnamon, salt, and pepper. She and her husband regularly make a giant pot, "and then we eat them with everything for days and days."

Like all heirlooms, the beans Arnold Stephano, Lubecki, and other farmers are putting in the ground in 2020 are the latest link in a long chain through history, connecting thousands of seasons of hard labor, sun and rain, and dirty hands that carefully saved seed year after year. "I'm not sure that my ancestors ate beans, they were from Norway—did they eat beans in Norway?" she asked with a laugh. "But it does connect me with my farming ancestors, everyone who has farmed through time." ○

Note

1. Sophie Ackoff, Andrew Bahrenburg, Lindsey Lusher Shute, "Building a Future with Farmers II," National Young Farmers Coalition, November 27, 2017.

Margaret Walker

LINEAGE

My grandmothers were strong.
They followed plows and bent to toil.
They moved through fields sowing seed.
They touched earth and grain grew.
They were full of sturdiness and singing.
My grandmothers were strong.

My grandmothers are full of memories
Smelling of soap and onions and wet clay
With veins rolling roughly over quick hands
They have many clean words to say.
My grandmothers were strong.
Why am I not as they?

Source: *This is My Century: New and Collected Poems* (University of Georgia Press, 1989)

Ungelbah Davila

PORTRAIT SERIES

Jessa Rae Growing Thunder

Ramona Emerson

Shaandiin Tome

Virgil Ortiz

↑ **Jean François Millet** *The Sower*
→ **Anonymous** *Rots in de wintertuin van de École Nationale Superieure d'Horticulture in Versailles*

Wilderness •
Compost

Frankie Gerraty

FROGS AND FUNGI

Caring for Microbial Worlds

> Life on earth is such a good story you cannot afford to miss the
> beginning.... Beneath our superficial differences we are all
> of us walking communities of bacteria. The world shimmers, a
> pointillist landscape made of tiny living beings.
>
> —Lynn Margulis and Dorion Sagan, *Microcosmos*

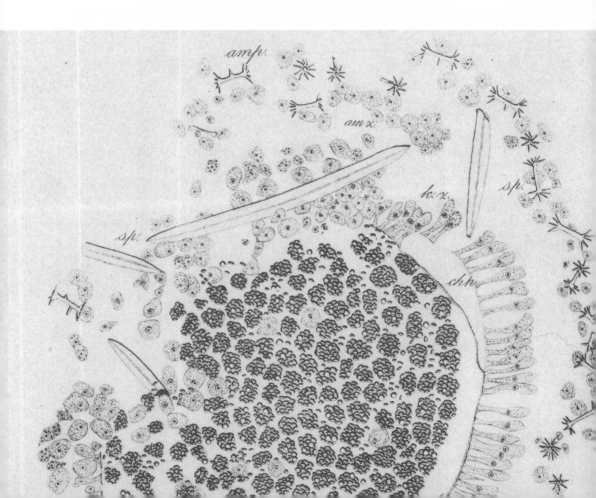

Shimmers of early morning sunlight illuminate miniature riffles on the surface of an alpine pond. Its icy water is held by cupped hands of granite, a glistening pool nestled into the floor of a glacial basin.

Beside the pond, a couple of stirring bodies shift in sleeping bags. As they wriggle from their insulated cocoons, the first freezing inhalations of the day tighten and chill their waking human chests.

The sun rises slowly in the Dusy Basin, in Kings Canyon National Park, and the collection of pond-side humans—a team of amphibian biologists—begin their work just as it starts to peer through the crown of granite peaks which rims the basin. They are conducting a survey of Sierra Nevada yellow-legged frogs, an endangered greenish-brown and yellow amphibian that inhabits bodies of water like this pool in the High Sierra. It is the summer of 2013, one of the driest in California's recent history.

They begin by walking slowly along the pond's edge, listening attentively. They hope to hear the rasping calls of breeding males and the plopping sounds of frog bodies retreating into the water in fear of approaching humans.

"I really love and care about amphibians," remarks Dr. Vance Vredenburg, who leads the frog survey team. Vredenburg knows this mountain range well. Starting in the mid-1990s, he began visiting each summer to survey dwindling populations of yellow-legged frogs.

. . . .

In 2005, Vredenburg's field season took a surprisingly lethal turn. His surveys revealed thousands of dead frogs littering the shorelines of his study lakes and ponds. The mass mortality left him heartbroken.

The frogs were diseased. When Vredenburg's collection of skin swabs were processed in the lab, he diagnosed the obliterated amphibians with a fungal infection called chytridiomycosis.

The microbial culprit behind the outbreak was *Batrachochytrium dendrobatidis*. Also known as B.d. or chytrid, *B. dendrobatidis* is an infectious fungus that likely originated in the Korean peninsula and has spread rapidly across the planet in recent decades. It was first noticed in the 1980s, first named and identified in the 1990s, and has since driven declines in over five hundred amphibian species on six continents. Chytrid outbreaks have led over ninety species to extinction.

B. dendrobatidis spores infect the moist and vulnerable skin of amphibians, a class of animals that includes frogs, toads, newts, salamanders, and obscure worm-like creatures called caecilians. Infection causes the skin to thicken and flake off. Because amphibians breathe through their skin, this thickening and flaking interferes with electrolyte transport and hydration, eventually leading to death via cardiac arrest.

Scientists have begun to call the chytridiomycosis outbreak an "amphibian apocalypse."

This panzootic disease has caused the greatest pathogen-driven loss of biodiversity ever recorded in human history.

Horrified by the devastation wreaked by chytridiomycosis, Vredenburg started to spend all of his time obsessing over the disease, its causes, and the rippling consequences of decimated frog populations on California's alpine ecosystems. He began to ask questions like, How does chytridiomycosis affect various frog populations differently? What are possible mechanisms for slowing and preventing chytrid's spread?

In the Sierra Nevada, the pathogen moved across the landscape at a predictable rate in a "wave" of infection. However, yellow-legged frogs near Yosemite fared far better through disease outbreaks than frogs in Sequoia and Kings Canyon National Parks, with some individuals surviving the infection. The cause of this discrepancy? A mystery.

Vredenburg began a series of experiments to help unravel the variation in frog mortality in different parts of the species' range. His experimental results were surprising; the primary differences between the semi-resistant frogs in Yosemite and the chytrid-annihilated frogs to the south were dissimilarities in the communities of symbiotic bacteria living on the amphibians' skins. Yellow-legged frogs in Yosemite were much more likely than their southern counterparts to carry *Janthinobacterium lividum*, a bacterium which exhibits anti-fungal properties. The researchers suspected that *Janthinobacterium lividum*, colloquially known as *J. liv.*, was helping frogs survive bouts of infection with the lethal fungus.

The ubiquity of this type of mutualistic living arrangement, in which an animal's microbial community helps to defend the animal's health, has become much more apparent with improvements in biotechnology equipment. These technologies have transformed how scientists understand both human and non-human bodies. Each of us—humans, other animals, fungi, and plants alike—are composed of riotous and bustling microbial worlds. Our bodily ecosystems teem with life, including diverse communities of bacteria, protists, archaea, and viruses. The health and balance of these microbial ecosystems is intimately linked to the health of the host organism.

For conservation-minded scientists like Vredenburg, this symbiotic understanding of life enables new techniques for sustaining the health of other species in times of emerging infectious diseases.

To defend the ongoing survival of his beloved frogs, Vredenburg brewed up vats of the purple-colored bacterium *J. liv.* with the hopes that he could use this bacterial culture to inoculate frogs in front of the epidemic's wave. Preliminary lab-based trials were successful, and he received permission from the National Park Service to conduct an experiment in a wild frog population in 2010.

The site of his experiment was in the Dusy Basin, home to a rapidly deteriorating population of yellow-legged frogs. Frogs once occupied several large lakes in this basin, but invasive trout forced frogs into the last remaining patch of suitable habitat—a small, shallow pond that dries out during exceptionally dry years.

Later that fall, Vredenburg's results suggested that his probiotics had successfully slowed the mortality caused by chytrid. After the wave of disease had passed through the Dusy Basin, the frogs he had inoculated with *J.liv.* earlier in the summer were surviving. The enduring frogs had still been infected by chytrid, but their levels of infection were much lower than the untreated frogs.

Three years later, however, Vredenburg's 2013 amphibian survey team completes their slow, attentive circumnavigation of the pond without hearing any raspy calls. Splashing frogs are nowhere to be seen. Sierra Nevada yellow-legged frogs have finally disappeared from the Dusy Basin. An eerie silence permeates the granite valley.

Vredenburg's brief conservation success had been only a small eddy of hope in the midst of a planet-wide cascade of amphibian loss.

Isaac Chellman, NPS   *Recovered mountain yellow-legged frog*

Diseases are not the only threat to frogs that humans have brought to the Sierra Nevada. Invasive trout, introduced via airplane drop for the purpose of recreational fishing, have radically altered fishless aquatic ecosystems across the Sierras since the late 1800s. These salmonids are voracious predators of mayflies and tadpoles, and consequently serve as a primary threat to the survival of alpine creatures including birds, such as grey-crowned rosy finches, and amphibians, like Vredenburg's yellow-legged frogs.

The threats posed by invasive trout and infectious pathogens are compounded by the hazards of changing weather and climate patterns in the Sierras. In the end, yellow-legged frogs in the Dusy Basin survived the onslaught of chytrid fungus with the help of their purple bacterial companions, but they perished as a result of predatory trout and extreme drought driven by a rapidly changing climate.

Conservation always happens in context, explains Vredenburg, and "the context of the Sierra Nevada is that humans have really changed the natural order of things." However, he hopes that negative human impacts can be matched with positive ones. From the careful manipulation of microbial ecosystems on frog skin to the removal of trout from alpine lakes, Vredenburg aspires to create a world in which humans cultivate ecologies of care.

Amphibians will persist, he believes, if we begin to devise new ways of caring for the spectacular beings with whom we share our planet. He is certain that "if we give these species a chance—just a small chance to survive—they will."

Turning our attention earthwards and beginning to notice the entanglements of frogs, fungi, bacteria, and birds is the necessary first step in moving towards this frog-filled future. ○

Emily Haefner

NETTLES

The earth becomes me.

The thistle and the nettle have tattooed my body.
My toes dig down into the soil.
The pondweed has tangled in my hair.
The cut grass dyes my robes emerald green.
I have eaten the wild herbs and the windowsill flowers.
I made salves from mud, clay, and sand to cover myself.

I am turning into a wild thing.
I race out into the night, howling,
drinking in the moonlight.
In the day I stand, opening,
taking energy from the sunlight.
I send my power radiating outward to hold fast as storms gather.

When you see me next season
I will be more deeply rooted.
My skin will be layered like aspen tree bark.
My head will hold nests of newly hatched songbirds.
Bees, butterflies, and other winged things will land on my fingertips
as I open my petal hands wide.
Around my feet will grow poppy and willow and sedge and clover
and thistle and nettle.

Earth becomes me.

Addison Bowe

Lauren August Betts

ATTEMPTED LOVE LETTER

(Almost) A Year at Full Belly Farm

Near the end of last year's rainy season, I went with Antonio—or as my brother calls him, "the hero of the farm"—to the pasture on the other side of Road 16 to visit the lambs and their mamas who were still living together outside the barn where they were born. On this farm, and I'm guessing on all others, places are always becoming other places and then turning back into themselves again.

When I arrived here in the middle of January, one of my first jobs was to weed around the small trees I now know to be olives, planted along the road leading past the new almond orchard to the barn. Another was to patch up tarps on a triangular structure where the lambs and moms could go to take cover when it rained.

At that point, the mamas were grazing the pasture outside of the barn, waiting for their moment. Now it was April, and they were back outside the barn—only they were each being followed through the grass by two or three babies.

It was my first time across the road in months. As far as I know, the shelter I patched was never used; the olive trees had since been swallowed by new weeds. In the barn, where so much had happened, there was now only straw and unfinished buckets of water.

On our way up in Antonio's little green truck, with its orange water jug attached to the outside and the right passenger door that only opened from the inside, I looked out onto the pasture on the property next to ours. The blades of grass were so tall and dense that when the wind blew through them they moved together as one big mass with ripples and waves. I was missing someone hard at the time, and although these thick blades of grass acting like water had nothing to do with him, watching them move that way made my stomach ache. Sometimes the earth is beautiful in a way that makes you remember everyone and everything you've ever loved.

Antonio and I drove past the new almond trees and past the shelter that had never been used and the olive trees we had weeded, then forgotten. We turned off the truck and waited for a moment, listening. When I'd first arrived, the other interns and I were told that when we pulled up at the barn to check on the lambs (usually in the middle of the night), we should turn off our truck and the lights and wait quietly for a while and just listen. Antonio and I did that this time too, even though there were no moms in labor or lambs being born to listen for.

Antonio got out of the truck and began to fill up the *tina de agua* for the sheep, and I found myself wandering into the barn.

I was looking for lumps of fur in the pens like the ones I used to see during lambing season, when I'd find one or two or three teensy-tiny babies curled up, trying to keep warm. The lighting was not great—the babies were mostly born or discovered late at night or early morning—and it was both incredibly beautiful and scary. Sometimes a little lamb didn't make it. When you saw one, you were looking at the possibility that it might not be moving. That never actually happened to me, but when I saw a lamb I would hold my breath and stare as hard as I could until I saw its body inflate and deflate, just a little.

I was looking now but there was nothing.

To explain how strange it is to watch a year go by on a farm feels impossible. When the seasons change, so many parts of the farm become ghosts of themselves; teeming and important one moment, empty the next. But they are just resting and waiting for the next year when they will be full again with their purpose.

Across from the peach orchard is a greenhouse where I spent a lot of time when I first got here. On what I'm pretty sure was one of the rainiest days in recorded history, I was tasked with saving seed from dried cayenne peppers. I was extremely excited because from the beginning I really liked monotonous solo tasks, especially ones in the greenhouse. I felt safe and frozen in time surrounded by white, listening to the rain hit the plastic. I sat on an upside down box and cut apart the peppers using those red-handled, long-nose snippers that seem good for pretty much everything, and dropped the seeds into an envelope, weeding out the ones that didn't look healthy. It was cold, but in the greenhouse it was warm. It was dark, but I didn't mind so much because I was falling in love.

Andrew, one of the six partners at Full Belly,

Rose Robinson *Charlotte: Birches*

came into the greenhouse after a few hours and found me deconstructing the chiles and pulling out the seeds bare-handed. Next thing I knew I was sitting at his kitchen table with all of my fingertips submerged in a bowl of milk. A piece of chocolate waited for me on a napkin next to the bowl for when the ordeal was over. I remember feeling like I was so, so new. That was the very beginning of a year that I spent learning so much that I thought my head was going to explode. I had never even seen a cayenne before. I hadn't thought about the fact that peppers are spicy. I wanted to touch everything there with my bare hands. Even at the end of my time at the farm, after a year of thinking, *hey, maybe if I put some gloves on before transplanting 40,000 onions, my hands might hurt a little less tomorrow,* I could never bring myself to put gloves on.

One of the most wonderful things about working here: sometimes you transplant so many onions that your hands smell like onions for weeks, no matter how many times you wash them.

I went to visit that greenhouse late in the summer and it was empty. I could stand at one end of the greenhouse and see right through to the peach orchard. Every year, it is filled with dried cayennes, and then it is filled with lettuce starts, and then onion starts, and then nothing. I mourned the peppers, and the lettuce and the onions and the rain. It didn't seem like there could ever be enough time, enough energy, enough peppers, enough water in the sky, to come back around.

Now the greenhouse is filled with dried cayenne peppers again. The rain is coming and it will be here for what seems like forever. And in January or February, it will be lambing season.

I wish I were better at writing love letters so I could tell every person on this farm how grateful I am and how much of an adventure I've had here. You are all outrageously wonderful and strong teachers, and it has been an honor to work with you to help the land along in growing the fruit and beans and corn and wheat and barley and almonds and walnuts and herbs and okra and pigs and chickens and goats and cows and everything else.

I said there was nothing in the barn when I went with Antonio to check on the retired mamas, but that's not entirely true. In one of the pens we found some fluffy little balls. A cat had repurposed the place where the sheep had their babies to be the place where she could have hers. It was perfect. Warm soft straw, tucked away from predators, clean water. We decided to return the next day, to check on the kittens. We came the next day—and the next and the next—and one day the mama cat was gone. The kittens hadn't moved. Antonio knelt down and pulled one up close to his face.

"Mira, los ojos. Todavía están cerrados," he said, looking at me with his eyebrows raised. *"Tenemos que abrirlos."*

I looked at him blankly, thinking about how thick and rough his fingers were. "Are you sure you can just do that?"

He nodded and began to pull open the kitten's eyes with his fingers. Once their eyes were open, they would leave, he said. He always surprised me with how gentle he could be. I'd thought those kittens were goners. They weren't moving at all, they had no food. Sure enough, with open eyes those kittens started squirming and making noises. The next day, they were gone. ○

Krisztina Mosdossy

ETHNOPEDOLOGY

I am increasingly intrigued by ethnopedology, the study of soil knowledge. *Ethno* refers to humans and *pedology* means soil science. Ethnopedological research has concentrated on the tropical populations of our planet and especially on cultures that have a spiritual connection to land. *Kosmos*, the spiritual aspect of the *corpus-praxis-kosmos* triad central to ethnopedological work,[1] refers to the spiritual connection to land which people incorporate in their land management practices. *Corpus* refers to information cognitively stored, and *praxis* is the practical application of information and spiritual knowledge to the land. Through the practice of living out this triad, people have regenerated soil for millennia, as evidenced by global indigenous practices dating back tens of thousands of years.[2] The increasingly familiar *terra preta*, or dark earth, is the result of the original biochar used by pre-Colombian people in the Amazon to build soil that lasts for millenia. From around 60,000 years ago—or, some argue, 120,000 years ago—the Aborigines in so-called Australia have recognized the finite nature of soil and practiced careful replenishment of what they hunt and fish.[3] Similar practices are revealed in the history of Native Americans in present-day California.[4] Regenerating soil is more common in our history than we are lead to believe.

That people have known how to regenerate soil for millennia, without the use of microscopes or litmus tests, is an incredibly liberating thought. This means that we already have the knowledge; the process of interacting with soil is conserved in our DNA. I look to my grandparents and great-grandparents, many of whom are and were gardeners. Their pedological knowledge was learned through storytelling by their parents and grandparents. I did not pick up on the importance of this knowledge until very recently. It's only in the last few years that I prioritized learning how to garden. I grew up in an "information age," and now the information and knowledge I most value is threatened by a virus targeting our elders. I am so lucky that my grandmother, Grandi as we call her, is alive and well today. She blesses me with wisdom. Early in the novel coronavirus outbreak, Grandi quoted from her favorite book, the Bible. She read 2 Chronicles 7:13–17. While this is a well-known parable shared in times of catastrophic suffering, this chapter emphasizes our need to take care of the earth, take care of the land, take care of our soil. Biblical ethnopedology—perhaps a new field in science!

We cannot care for our soil without feeling a connection to it. Kosmos, the spiritual aspect of soil knowledge, is rooted in culture, symbolism, and ritual. Many cultures around the world continue to refer to symbols and practice rituals that maintain their connection to land. Some of us searched long and hard to have the land remind us of an ancient language. Our ancestors needed to know the land intimately to survive, a connection that evolved as part of the human spirit. I find myself wondering more and more if, through ethnopedology, our land-based ancestors sensed the microbial activity in the soil. Could they detect the predator-prey dynamics between nematodes and bacteria? Could they feel the tentacle-like mycorrhizal hyphae extend beyond a plant's roots and exchange nutrients deep in the soil with the plant? I think our ancestors were at least aware of the human role in maintaining soil's complexity. I also think there are people in each of our communities with magical abilities who can detect the soil's nuances. Perhaps their magical powers tell a similar story to that of soil science.

The very nature of soil, the foundations of all land-based life, is living in community. Like a human community in our immediate environment, soils are infinitely complex; bacteria and fungi dominate the microscopic scene, dodging nematodes and protists who are hunted by the mighty earthworm and oribatid mite. As the most populous in many soils, both in numbers of individuals and numbers of species, bacteria and fungi often fulfill similar functions, rendering some species redundant. How wonderfully resilient! If one species is lost for whatever reason—drought, tillage, predation—another species might survive and fulfill a similar role. This way, the system continues providing usable nutrients for the plants above ground. Plants depend on us to ensure we do not damage the soil so much that resiliency is

lost. This resiliency is further threatened by the fact that soil science is an understudied discipline. The majority of species found in the soil are unknown to science and are not artificially reproducible—we cannot culture them in laboratories. So, while soil-dwelling species may be able to fulfill one another's roles and continue to reproduce and maintain their own life cycles, they cannot do so without soil. Perhaps the purpose of ethnopedology is to recognize the sacredness behind the soil so that we prioritize maintaining its resiliency against known threats, such as climate change. Ethnopedology serves to remind us that soil is irreplaceable.

Moving forward, I intend to foster my own sense of spirituality with the land and pass on my ethnopedological knowledge to the next generation (and anyone who will listen!). I hope to follow in the footsteps of educators and scientists who are working to bridge the gap between science and human connection to land and prevent the world's soil supply from disappearing within the next two generations. If we gradually reinstate soil as the foundation of our humanity through ethnopedology, regardless of where we come from or if we own land, we can fulfill our responsibility to the land and future generations. ○

Notes

1. N. Barrera-Bassols and J.A. Zinck, "Ethnopedology: a worldwide view on the soil knowledge of local people," *Geoderma*, 111, no. 3-4 (2003): 171-195.
2. See John Russell Smith, *Tree crops: A Permanent Agriculture* (Washington, D.C., Island Press, 2013); Susan-Jane Beers, *Jamu: The Ancient Indonesian Art of Herbal Healing* (Vermont, Tuttle Publishing, 2012); Bruno Glaser and J. J. Birk, "State of the scientific knowledge on properties and genesis of Anthropo-

genic Dark Earths in Central Amazonia (terra preta de índio)," *Geochimica et Cosmochimica Acta* 82, (April 1, 2010): 39-51.
3. Bruce Pascoe, *Dark Emu: Aboriginal Australia and the birth of agriculture* (Australia, Magabala Books, 2018).
4. Kat M. Anderson, *Tending the Wild: Native American Knowledge and the Management of California's Natural Resources* (California, University of California Press, 2013).

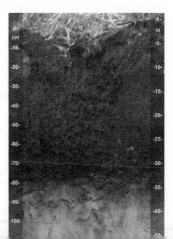

Lauris Phillips

THE LEAN, UNISON CALL, COURTSHIP

My work with Japanese sumi ink is both a pathway into and a reflection of an intimate relationship with birds. Painting with ink is as old as writing and evolved into a Zen art in China and Japan. The energy of life, of the painter and painted, is the actual subject matter and is only conveyed when the painter and painted are no longer two.

For five years, I have been photographing and painting Greater Sandhill Cranes in northeastern California, where they migrate to nest. At the Modoc National Wildlife Refuge, I assist on a long-term study aimed at protecting this vulnerable population. In that time, I have been enchanted with the variety of expression and communication within their "culture." They live for a long time, mate for life, and have been on the planet since the dinosaurs. They are complex and intelligent and huge. And they dance.

In the spirit of non-harm, my tools are made from found fur and other scavenged natural materials, and the ink I use is non-toxic, made from soot, as it has been for centuries. ○

← *The Lean*
Before flight, one crane leans forward and glances around to see if there is consensus. Others might lean and then stand up again, cancelling the flight, or lean further in agreement. The lean increases until they break into long strides, unfold their seven-foot wingspans, and run into the air.

↑ *Unison Call*
When bonding or waking, or if they hear another pair in the distance, cranes call in a resonant duet of sound that energizes the marsh.

→ *Courtship*
Each spring, when migratory flocks are on the move, young cranes dance passionately for each other with elaborate dips and twirls. Lifelong partners court as well, using the same innate gestures throughout the mating season.

Ang Roell

RADICALIZE THE HIVE

What Honeybees Can Teach Us About Social Change

Originally I wanted to write about carbon drawdown, planting for pollinators, and the intersections of soil health, human health, and honeybee health. But in the midst of a public health crisis with no known end, I couldn't bring myself to say "planting flowers" is our cure-all.

Uncertain what to do, I went to "tell the bees" about COVID-19, social solidarity, and physical distancing. Telling the bees is a ritual of bringing grief and sorrow to the beehive to share. It is said that if a beekeeper dies and no one "tells the bees," they will die too.

In trying to explain COVID, I told them, "If we came together now, we could get sick, and many of us could even die."

They responded, "If we were apart, and could not hold each other, even for a short while, we would most certainly die."

In times when we need community solidarity, thoughtful action, and mutual aid, it feels relevant to practice remembering. How can humans begin to build new ways of being in community with one another by remembering that we are, in fact, reliant on each other for survival? For me, because of my own agricultural practice, this remembering begins with the honeybee.

The honeybee-human relationship is old. There is evidence that humans were gathering honey ten to fifteen thousand years ago, before they had begun to farm or domesticate animals. An eight-thousand-year-old rock painting discovered at Arana Cave near Valencia, Spain, depicts a person climbing a ladder to gather honey from a hive on a cliff face. Perhaps these foragers could be considered the first bee "keepers."

Today bees are used as a tool for pollination in North America, but beyond its agricultural function, beekeeping forms, facilitates, and enables an esoteric, physical, emotional relationship between humans and bees. The relationship bridges cultures and continents. It's love—ritualized, interspecies affection. It queers time, gender, and nature.

In contemporary beekeeping, the goal of many beekeepers is to create apicultural systems that thrive. Beekeepers want to manage systems that are healthy and well. As longtime beekeeper Kirk Webster once told me, "Farming as a system is a series of compromises." Compromises are required because farmers are trying to manage the wild, "tame" the land, and "produce" an end product.

Turning towards human interconnectedness with this question, I wonder, what are the compromises we've made in our social systems? Where are our negotiations taking place? Who

are they impacting?

Our current economic model is based on endless growth, conspicuous consumption, and a perpetual race for profit. This global economic system threatens the ecological balance of our planet and has multiplied inequities. As humans participating in this system, we stand on the precipice of ecological crisis. If we cannot imagine systemic change, we cannot begin to end the extractive practices that have perpetuated the oppression of people, land, water, and animals since the so-called Age of Exploration that precipitated the arrival of Europeans in what became known as North America. In the year 2020, with the COVID crisis and police violence in the United States impacting Black and Indigenous communities disproportionately, and racism becoming a central topic in the American consciousness, it is difficult to deny that each of us exists within the framework of systemic oppression.

With practice, we can actively examine how oppressive patterns show up in our minds and in our daily lives. Patterns of fear and shame take root inside of us. If we're not actively doing the work to build awareness and deconstruct these patterns through healing and equity building, we are not doing the work of radicalizing or abolishing the systems of oppression.

But we can! We can shape change if we remember Audre Lourde's well-known declaration that "the master's tools will never dismantle the master's house."

We need to use our courage to address our collective needs by turning to our natural and elemental allies and ask them, what can we learn? What can we humans and nonhuman allies be together? What can we do together?

The earth and her fauna and flora have much to teach us. We need to remember that we are a part of a place, a history, and a universe much bigger than ourselves. When I listen to the bees, I hear the stories of an organism—the honeybee hive, ready to teach us about the dangers of industrialized expansion and the power of successful cooperation.

In contrast to the profit-driven ideals of industriousness and productivity, honeybees are translators of sweetness and light. They are facilitators of pollination. We can apply honeybees' role in the ecosystem to our work to build networks of collective care and shared power. True collaboration is sharing power with each other rather than holding and manipulating power over one another, or over another species. This is how I have shaped my iterative practice and work, and here, I share that with you.

Lessons from the Honeybees

Activist and facilitator adrienne maree brown says, "Small is all." When we start small, we can start by reflecting on our visible and invisible social identities, including race, ethnicity, class, gender, sexuality, and ability. We can notice and name where each of us holds power and privilege, because you can't share power if you don't know you have it.

You can address big issues by starting with small actions to combat injustice in the small groups and communities you're a part of today, and grow your practice slowly over time. The lessons we can learn from the hive are lessons on iterative process. In a healthy hive, we see some basic essential functions and lessons we can adapt to our own groups from each.

1. Essential Function

Mechanisms for clear communication are built into the biology of a honeybee hive. In a hive, individuals work in roles, practiced with ritualistic precision by each bee, for the collective benefit of the hive's longevity. Every hive is made up of more than thirty thousand worker bees, one queen, and several drone bees.

Lesson

In human society, diverse perspectives and identities are important, but they can only become powerful when we make space for them to be heard, valued, and made integral to the collective. Because we're not bees, we have to make space for humanity, which means making space for our differences *and* our traumas. We have to make room for healing and repair. The decision making within our communities has to be accessible for those marginalized by our current system. It has to call in our survivors and create space for compassionate listening and the bravery to share diverse perspectives and experiences as a group.

When we build high-trust groups like this, we can be coherent and effective. Bees trust innately; it is part of their biology. But we humans have to build it, just like we build new neural pathways in our brain when we learn new habits. Then we can be agile and adaptive in a complex and rapidly changing environment, while sharing power.

2. Essential Function

In a balanced ecosystem, honeybees are facilitators of pollination—transforming sunlight into sweetness by transmuting nectar and pollen into resources for their young. They are facilitators of interspecies sex, humming between stamens and pistils. When hives are gathering these important resources from nectar-producing plants, they are full up with pleasure and drunk on sunshine. They are in a reciprocal relationship with the ecosystem in which they are pollinating. They draw resources from and contribute to the plants' capacity to thrive. Honeybees pollinate food they'll never eat, store nectar and pollen for young they'll never meet, and swarm to locations they decide upon while suspended in midair.

Lesson

Our relationship with our surroundings can be joyful, purpose-filled, and inter-dependent. Humans can connect to our surroundings by slowing down, spending time in nature, connecting to the ecosystem around us. We can contribute resources to our communities and environment and practice reciprocity in our local bioregions.

3. Essential Function

Worker bees are the force of collaboration in a honeybee hive. Workers begin as cleaner and nurse bees who prepare wax cells for their sisters' births by cleaning, polishing, and adding food to each cell. They care for the young bees of the hive. Young worker bees also spend time as builders. They engage in an act called fes-tooning, in which they hang off of one another, secreting wax from scales on their bellies, passing it arm to arm to mouth. Then they chew the wax into hexagons in which to store young bees, food, and water. Foragers gather resources for the hive. They can travel up to eight miles to collect pollen, nectar, propolis, and water that will feed and hydrate the hive.

Lesson

Humans, like worker bees, play many roles and move between them fluidly.

4. Essential Function

Wax is the foundation of a healthy hive. It is the architecture upon which the entire hive is built, the central nervous system of the hive. Geometrically, hexagons hold the most weight with the least amount of material.

Lesson

We humans can work together to build solid foundations on which to collaborate.

5. Essential Function

Pollen and nectar are collected from flowers by forager bees. These resources are transmuted into food, called bee bread, for the young, and the surplus is stored for the next generation in the form of honey.

Lesson

If humans can learn to build together with an eye on the future, we can share in abundance and store it for moments of contraction.

Thomas Halfmann *Like Bee Hives*

6. Essential Function
Propolis is created from resins that
forager bees collect from evergreen trees.
Mixed with enzymes in the honeybee gut,
the resins become propolis. Propolis seals
the inside of the hive, making it a sterile
space for the young and protecting the
hive from moisture, cold, and predatory
insects.

Lesson
Humans can cultivate a space where it is
healthy for everyone to thrive.

7. Essential Function
One of the most crucial elements of a
healthy hive is clear, consistent, and collaborative communication. Honeybees
communicate through several methods:
vibration, dance, consensus building,
and pheromones, which are a form of
scent-based communication made
through glands on the honeybees' heads
and butts.

Bees take time to build communication that is iterative, equitable,
interdependent and accountable. They

building can help us recognize we're part of a larger whole.

Accountable communication in groups, organizations, and communities happens when we start small and build clear mechanisms for communication and accountability. Then we can aim to be accessible and build trust across differences, but this can only happen with radical honesty and iterative practice. Trust built when we aim for accountable and nonhierarchical systems, and this trust helps us build our capacity to be interdependent.

Bees are a whole world that can open us up to others—blossoming trees, spiders, ants, predators, opossums, and bears—and to how all of these things interact with each other. If we look to the ecological world all around us, to our allies in the plant, animal, and insect world, we can draw inspiration for how to shape change. We can tune into natural rhythms and lessons once again. We can take lessons from inside the hive to begin the process of building resilient worlds together beyond the hive. We can build collective care through shared networks.

When we can see systemically, in patterns, in iterations, in fractals, we realize how flawed our social systems and structures are because they are both limited and disconnected. The more we realize that, the easier it becomes to decouple from degenerative practice and align ourselves with an ecologically responsive, adaptive, and generative process. ○

eliminate hierarchy in their systems by communicating unilaterally about all of their choices, so they can make the best choices available to them.

Lesson

With practice, our human communities can become agile like honeybees. We can minimize hierarchy in our own social structures and build intimacy in our relationships that truly takes everyone into consideration and makes room for new ideas. Authentic human consensus-

Kaitlin Bryson and Hollis Moore

ITS VITALITY COMES THROUGH FLUCTUATION

The current stand of the most prolific cottonwood trees in the Middle Rio Grande Bosque are nearing the end of their lives. Cottonwoods grow when flooding occurs along the riparian zone, allowing for successful germination of seeds. With the rise of climate change and diversions upstream, there has not been a successful year of natural propagation since 1941. The decline of cottonwoods along the Rio Grande will change the entire ecosystem and habitat, resulting in species migration and loss. The structure of the Bosque will take on a new composition, increasing aridity and exposure, advancing the desertification of the Southwestern ecology.

The mural was painted with soil collected from the riverbed, black walnut ink, and ink made from cottonwood chlorophyll. It depicts the topography of the Rio Grande floodplain, overlaid with the human-built river diversion infrastructure. Cottonwood seedlings are painted in a line at the height that flooding needs to

occur in order for the seeds to germinate. The ink made from chlorophyll decays with light exposure, and as the exhibition continues, the painted seedlings fade.

Simultaneously, paper made from cottonwood seed pods and embedded with native plains seeds grows over a labyrinth of native animal tracks. If the cottonwood canopy disappears, the entire ecosystem will change and habitats for those animals will be lost. ○

Zoë Fuller

EPIDEMIC × 2

COVID-19 is not far from my mind as we walk our logging trail.

Nicky has been out in the woods all morning, tromping with snowshoes, hauling a sled that's a wooden box fastened to a set of old red skis, its runners. It's what we use to haul jugs of water to our home and he's repurposed it for logging. "Woodsing," we call it, rather than logging, which sounds so industrial. Woodsing means we're clearing a trail, getting firewood, or moving cut boughs and rotten stumps to decompose in a mound, eventually creating good growing soil.

Today we're hauling poles to make the farm's washstand, and we talk while we walk through dead and dying spruce. Their branches are brittle and their sap bleeds out, forming sticky mounds on alligator-skin trunks. We are small animals to these trees, one hundred years or more older than us. One was born in 1799—we know by counting its rings with a pencil point. Nicky says, "Who are we to cut them down?" I suggest that we sing to them, give them last rites worthy of the life they've lived, but the only tune we can come up with is "Twinkle, Twinkle, Little Star," which doesn't feel quite proper, but doesn't stop us.

We haul ten-foot sections back to the barn, hooking the homemade sled to a snow machine (Alaskan parlance for snowmobile) that barely works. The logs are heavy, their trunks still holding water as they did when they were alive. We used the tinder-dry trees for this year's firewood. These trees gave us wood to build our home and fuel to heat it. I have a list of build-ings our working farm needs—this washstand, an animal barn, a greenhouse, a root cellar, a cabin for when people come to stay, a tool shed, another tool shed.

We stand in the sun and lean sections of log up against one other, peeling their brittle, chewed bark with drawknives. Their two handles stick out at right angles with a blade between them that is only sharp on one side. As we peel, we see the black bodies of the beetles, their larvae curled into tunnels. Usually bark beetles are satisfied eating the cambium of old downed and diseased trees, but spruce bark beetles have multiplied exponentially in these last warm winters of south-central Alaska. They're chewing away at healthy spruce, killing tall trees early, so that we are living in a forest of firewood—standing dead, ready to burst into flame.

It's easy to vilify the spruce bark beetles, and I have seen men take delight in burning bark off stacks of downed trees with a propane torch, reducing spruce bark to ash, killing the beetles and their habitat with a pass of a flame. It's logical to slow the beetles' epidemic spread, and I understand the impulse to annihilate the imposer. I too have resented the insects for prematurely killing the trees that could heat our homes, build our structures. I have grieved the loss of the forest as I know it, but as I peel logs, I can't help but talk to the beetles I strip from their burrows.

They say, "We're breaking it all down as we chew spruce bark. We are the stewards, shifting our forests from one climate to the next. The

Étienne Léopold Trouvelot *Aurora Borealis: As observed March 1, 1872, at 9h. 25m. P.M.*

temperatures rise, human. You caused it, but it is so much bigger than your tiny lives."

I draw my knife against the bark. Left standing, and in the absence of fire, a red fungus will begin to eat at the dead trees' centers, decomposing them, turning wood to soil. The rotting logs seep nutrients back into the forest, their trunks staying damp with moisture during much of these hot, climate-changed subarctic summers. The bodies of the spruce trees will nourish what grows in our new climate. But logs decomposing with mycelium aren't what we need for building projects; the time to

harvest these big, glorious, dead spruce trees is this year, maybe next. We work in the woods with an edge of hurry and all the humility we know how to give.

*   *   *

I am a farmer living in the woods, and during the "stay-at-home" era, my life stays much the same. I start seedlings for the farm under fluorescent lights to be planted when the soil thaws. I teleconference, call old friends, and worry for others across the globe. I feel a tightness in my chest and it's not sickness, but panic for the ways this pandemic widens systemic inequality.

I worry for the woman who works at the coffee shop, for my friend who cuts hair for a living.

On a walk, we see a face we recognize, a neighbor on a four-wheeler, her sheepdog bounding near the wheels. I wave and she recognizes me. We speak of garden plans, the conditions of the snowpack, how spring might be late this year; it's melting so slow. Our dog leaps into her lap and licks her ear, unmasking our veneer of casualness. To meet another on the path! To speak of the weather, safely six feet away. Visiting has become sacred, risky. We must trust in another's consideration—not to sneeze, not to reach for a hug. Not to break the barriers of space that are our protection from a virus that so far has not been reported in Alaska.

It's been seventeen days of near-quarantine, of slowing down, and I feel that I might have forgotten how to drive a car. There has come a stillness to my life—a stillness for many of us, though I am deeply aware of the ways that this pause is painful for many in the world. In the United States, some strike, demanding that rent payments be halted. Shelter in place, they say, but what of the money that buys shelter? Will the people rise up by staying still, and say, "Our homes are where we live, and no, we will not give you money and we will not move"?

Sixty miles east of the eastern border of Alaska, the Gwich'in community of Old Crow had to account for two uninvited visitors from the city of Quebec. They came seeking refuge from the pandemic, intending to make a new life for themselves off the land, coming with no tools and no gloves. What they brought was potential transmission of the virus to a people with limited access to health care, vulnerable elders, and the tremendous weight of a history of pandemics that decimated and traumatized their communities. Members of the Tribal government met them at the airport, proactively isolated them before they could transmit the virus, and sent them back.

Nicky's mother calls us. We hold up the phone to show our dog bounding through snow-covered farm fields. She lives in Anchorage, loves to ride her bike to a bakery each day, and works from a downtown office. She says, "In our very last resort, can we come live with you? I'll work on the farm, do lots of shoveling." It was said with the air of a joke that wasn't, and she says, too, "Even if we can't drive cars anymore, we'll walk there."

In the petroleum-producing state of Alaska, the price-per-barrel of oil has always been the metric of prosperity, printed in news, stated on the radio, in conversation. I've heard discussion of this metric since I was a child. The number is always in flux, celebrated when it's high, propped up when it's low, a dismal measure of well-being. I imagine buying a barrel of oil myself: Is it like the barrels I use to hold rainwater? Can I use the crumpled bills in my pocket to pay? It's down to twenty-eight dollars. I can afford that, but I'd rather have a barrel of compost. All around me, people turn to gardening. The places that sell seeds are running out, as are those who sell chicks, ducklings, the start of a laying flock. I recognize the urge to localize, to have eggs and vegetables if the grocery store doesn't. It borders on hoarding, but I see, too, how humans want to care, to cultivate, to be closer to the earth.

This virus invites itself into our home, saying, "I will show you how you push each other to slow deaths, humans. I am both the creation of your exploitation and the messenger of the earth. I will grind the gears until you stop. You will look up at a sky with no airplanes in it. Slow down, walk in the woods or wherever you are, and learn to live in reciprocity." ○

Matsuo Bashō
Translated by Lafcadio Hearne

Old pond – frogs jumped in – sound of water

Alex Hiam
Pintail Passing

↑ *Bees* from *The Home and School Reference Work*, Volume I
by The Home and School Education Society
→ **Jack Delano** *Yabucoa, Puerto Rico. Sugar strikers picketing a
sugar plantation*

Arguments

Conservation Corps, the White House

Elena Vanasse Torres

A BRIEF HISTORY OF LAND TENURE AND THE JÍBARO CONDITION ON THE ISLAND OF PUERTO RICO

It was the hot summer of 1898 on the southern coast of Guánica when 1,300 North American[1] soldiers first stepped foot on the island of Puerto Rico. *Weather delightful*, wrote General Nelson Miles to the United States Secretary of War in Washington, DC. *This is a prosperous and beautiful country. The Army will soon be in mountain region.*

As Miles,[2] lieutenants, and soldiers disembarked their ships, the turbulence of seafaring behind them, their sweethearts safe at home, they stepped into another dimension, where papayas hung sensually and mango meat melted off its seed in the heat. They licked their lips of the ocean's salt and ventured inland, where the mountains rang with tree frog hymns, and the weather was, notably, delightful.

In the following months, the United States would face a series of clashes with the Spanish, who had preceded them by over four hundred years in their reign over the island. When the US military worked its way into the mountains of Adjuntas and of Utuado, they were met, peculiarly, by parades, fireworks, ringing bells, and cheering *criollos*,[3] who believed that the North Americans had come to liberate them from the chokehold of Spanish rule and would herald the sovereignty they so desired.[4]

The rural workforce, especially plantation owners, perceived the arrival as an opportunity to establish a line of commerce within lucrative US markets—potentially, a free market system that would also guarantee them earnings in dollars, not pesos. These aspirations were realized upon the US's formal occupation of the island a few months after their arrival, but a gross corporate takeover of Puerto Rican lands followed. Some plantation owners sold their land, others leased to the North American *centrales* where sugarcane was ground for processing and the owners were able to do business as absentees, sitting fat and rich in their villas in Palma de Mallorca, Barcelona, and Paris. In the words of Luis Muñoz Rivera:

Los productos de la campiña feraz toman, año tras año, el camino de Europa; en que los ingenios permanecen junto al Portugués, y el Inabón y al Bucaná, pero las rentas se consumen en Paris, en Londres, en Madrid, o en Barcelona; en que los capitales emigran y acuden a fomentar el lujo de otras ciudades en que se alzan palacios y jardines que llevan en sus sillares y en sus invernáculos el jugo amargo de nuestros cafetos y el dulce jugo de nuestras cañas.

The products of the fertile countryside take, year after year, the path to Europe; in which the plantations remain by the Portugués,

Inabón, and Bucaná rivers, but the profits are consumed in Paris, in London, in Madrid, or in Barcelona; in which the income emigrates and comes to promote the luxuries of other cities where palaces are erected and gardens, embedded in their ashlars and greenhouses, carry the bitter juice of our coffee trees and the sweet juice of our cane.[5]

As under Spanish rule, the wealth of the mill owners was built on the backs of the sugarcane proletariat, who endured long, grueling workdays, backbreaking labor, and the seasonal nature of the *zafra*,[6] which left workers idle for a significant part of the year. The majority of Puerto Ricans at this time were illiterate, landless agricultural and subsistence farmers who relied on measly daily wages to support their families.

At the time of the North American invasion, the Puerto Rican countryside was largely owned by the Puerto Rican and Spanish elite and, due to the decline in the sugar industry under Spanish rule, was characterized by vast expanses of underutilized properties; some former sugar lands had turned to swaths of grassy, fallow land during the sugar crisis of 1870. Interspersed were *agregados* (campesinos living on titled estates who often exchanged labor for usufruct rights to small parcels), families engaged in sharecropping, and informal settlements. These are some of the lands that the United States would snatch up to establish the centrales—Central Aguirre, Fajardo, South Porto Rico, and United Porto Rico—that by the 1930s accounted for nearly half of all the cane ground in Puerto Rico. By rapidly displacing agregados in the southern, cane-producing regions, the United Statesian occupation drastically altered[7] the patchwork of subsistence farming throughout the island. Any food, whether imported or homegrown, would become a commodity to be purchased with the *jíbaro's* measly wages. This disruption of a formerly insular rural economic model marked the beginning of a legacy of dependence on food imports[8] that continues to this day.

A new period in Puerto Rican politics began when the Partido Popular Democrático rose in popularity, posturing itself as the party dedicated to the rural worker's plight. By 1941, the *Populares* had implemented what became known as the Land Reform Law,[9] which sought to downsize the sugar *latifundia* and support domestic food production. The few successes of the reform, such as the *parcelas* program,[10] which sought to provide the rural working class with small parcels of land for subsistence, have been overshadowed by the fact that the reform failed to break up the vast expanses of sugar croplands, failed to change the deep inequality in land tenure, failed to create a movement of independent farmers or cooperative agriculture, failed even to diversify food production or reduce the slowly increasing amount of imported foodstuffs.[11] Although the parcelas program is touted as a success of the reform, it merely perpetuated the dependence of the rural classes on tenuous government programs. In Vivian Carro-Figueroa's words, "the agrarian reform, therefore, did not create a class of independent smallholders, but rather one totally dependent on the government's Land Authority for a place to live and work."[12]

In the wake of the Second World War, Operation Bootstrap was implemented by then-congressman Luis Muñoz Marín in consultation with the US government, allegedly to rescue rural people from their plight. In reality, the operation was designed to move the Puerto Rican economy away from agriculture and towards industry to meet the North American demand for material goods during and after

→ **Jack Delano** *Farm Security Administration borrowers and their families at a meeting held to discuss the distribution of land for tenant purchase farms*

World War II.[13] The agricultural economy was transformed to a modern, industrial economy practically overnight. Lured by the promise of a better life in the urban centers of San Juan or in the continental US, Puerto Ricans rapidly left the countryside. Between 1945 and 1964, close to 750,000 Puerto Ricans—more than one-third of the population—left the island.

Largely due to a decline in competitiveness in the global market coupled with wide-scale land degradation, sugar production eventually came to a halt[14] in the 1950s. Combined with the industrialization project, this caused agricultural employment to be cut in half. Today, only two percent of employed people in Puerto Rico are working in the agricultural sector, and not even half a percent of all employed women work in farming.[15] It can be said that today the campo is, generally speaking, impoverished, sparsely populated, and bereft of opportunity.

Further exacerbating these problems, most agricultural incentives available to Puerto Rican farmers today privilege large-scale monocropping and agribusiness,[16] disregarding the fact that most food crops for both subsistence and local commerce—coffee, banana, plantain, cassava, and yam—are produced by small-scale farmers in plots averaging fewer than three *cuerdas*.[17] It continues to be the case that rural people remain deeply disadvantaged socioeconomically, with little opportunity to gain education or training to jump through the bureaucratic hoops associated with applying for USDA or Land Authority grants and subsidies. They are often ineligible for crop insurance or agricultural loans; legislators and Land Authority officials don't even care to pretend that these opportunities are appropriate for the majority of rural people.

What can only be more painful than the rusting, dilapidated vestiges of the *azucareras* in Puerto Rico's southern coast is the decades-long administrative support of tax exemptions and subsidies that fail to protect small-scale

Jack Delano *Barranquitas (vicinity), Puerto Rico. Hands of an old woman working in a tobacco field*

Boricua[18] farmers and instead benefit multi-national corporations and high-net-worth[19] United Statesians.

Just under eighteen kilometers from where the US infantry disembarked in 1898 sits a Bayer experimental facility that, along with Monsanto, DuPont Pioneer, and five other multinational producers of GMO seeds, received more than $526 million in subsidies from the Puerto Rico Department of Agriculture and the Compañía de Fomento Industrial[20] between 2006 and 2015. During the same years, Monsanto and Pioneer were authorized to extract over 238 million gallons of water from pristine southern aquifers at no cost. Today, the *semilleras* claim 10,000 cuerdas of both public and private land, and cover nearly a third of prime agricultural lands in the Juana Díaz vicinity. As in the age of the sugar latifundia of the Northern state, Monsanto, Syngenta Seeds, Dow AgroSciences, Mycogen Seeds, DuPont Pioneer, Bayer Crop Science, and AgReliant Genetics all violate the 500-acre limit.

Meanwhile, 85 percent of all food consumed on the island is imported to satisfy the *new* average Puerto Rican diet. A diet that historically persisted as mostly home-grown produce and farinaceous crops with a few supplemental imports shifted decidedly to imported, pre-packaged, and processed foodstuffs following the introduction of the US Food Stamp Program in 1974. Being that farm wages are not exactly competitive with livelihoods centered around food stamps and other anti-poverty programs,[21] the vast majority of rural Puerto Ricans utilize *cupones*[22] to make ends meet. The island's chronic food insecurity was made even more apparent in the aftermath of Hurricanes Irma and María, when 90 percent of Puerto Rican crops were decimated, and logistical problems at the ports held up routine imports as well as vital food aid. These combined factors caused a widespread food shortage that quickly escalated to a full-blown humanitarian crisis. Amid seemingly uncountable deaths, catastrophic flooding and landslides, and a lack of clean drinking water, Puerto Rico's agroecological farmers picked up FEMA's[23] and local authorities' slack.

Before the hurricanes, grassroots *brigadas*[24] were already taking place since the 1980s. But when María and Irma decimated so much of the countryside, these brigadas were in full effect, cutting new paths, installing solar photovoltaic systems, rebuilding, replowing, reterracing, replanting, and putting quality heirloom seeds in the hands of farmers. The work of Boricuá, Colmena Cimarrona, Departamento de la Comida, Proyecto Agroecológico El Josco Bravo, Finca Escolar Agricultura en Armonía con el Ambiente, Plenitud PR, Siembra Tres Vidas, and Camp Tabonuco, along with the University of Puerto Rico's Mayagüez and Utuado campuses' Horticulture and Farming Programs, is activating the campo. Several, if not all, of these projects leverage their ties to urban and academic centers to acquire outside funding that supports the bulk of their activities, and, in some cases, their land payments. The need for fewer barriers to land ownership and stewardship, and more investment in projects that reconnect with traditional, agroecological means of subsistence farming, is clear.

What kind of land reform do we need to transcend a deplorable history of oppression and make land ownership more accessible to the rural population, as well as the growing number of young agrarians on the island? If we cannot even rely on the apparatus of governance to protect our lands and water, nor the right to caretake the lands our ancestors tended, to whom do we turn?

Implementing a development plan that places agrarian reform at the forefront, that increases credit availability for small-and medium-sized farmers as well as protections for domestic agricultural products, is an important step to revitalize rural Puerto Rico. The path to radical land reform, however—where fallow lands are redistributed, as has historically been done in Cuba[25] and Guatemala,[26] and attempted in Brazil[27]—would bring our island's most fertile lands, those that now belong to multinational corporations and the Land Authority, back into our hands. In imagining an island nation where free healthcare protects the people growing our food and free, quality education is accessible to people in the farthest reaches of the island—where rural peoples procure their own *conscientizaçao*[28] and have a seat at the policy-making table—we must consider the exit of the fiscal control board[29] and independence from the United States. We must finally liberate ourselves from

Notes

1. A note on terminology: North American is used in this text interchangeably with United Statesian. The term "American" is problematic due to a history of British-American/United Statesian imperialism and thus distortion of geography, where American no longer refers to the citizens of the thirty-five North and South American countries that make up the American continent. Although North Americans may refer to Mexican, Canadian, Puerto Ricans, or United Statesians alike, the term here refers to the North American infantry soldiers of the United States Armed Forces who invaded Puerto Rico by land on July 25, 1898.

2. Not surprisingly, Miles was a key player in nearly all of the US Army's campaigns against the American Indian peoples of the Great Plains.

3. Criollos (people of sole or mostly Spanish descent born in Puerto Rico) made up most of the rural working class at this time.

4. A few important caveats to this are represented in the actions of the legendary Macheteros in the Battle of Asomante and Aguila Blanca's revolutionary activities in Jayuya, Juana Díaz, and Ponce. And starting just six weeks after the establishment of the US military government, cane fields were periodically set on fire in protest of poor working conditions and meager wages—issues exacerbated by a 40 percent increase in the cost of living due to the new (US) currency, among other factors.

See Andrés Ramos Mattei, *La Sociedad del Azúcar en Puerto Rico* (Rio Piedras: Universidad de Puerto Rico, 1988).

5. Luis Muñoz Rivera, *Obras Completas: Prosa: enero a diciembre de 1893* (San Juan: Instituto de Cultura Puertorriqueña, 1960), 119-133.

6. Sugarcane harvest marked by influx and later efflux of farmworkers from various regions of Puerto Rico.

7. Landlessness, at least on paper, actually declined between 1899 and 1910, possibly in response to the Hollander tax on land in 1900. According to scholars Cesar J. Ayala and Laird W. Bergad, the new land tax placed pressures on the largest landowners with idle lands to put them into productive use or to sell them: "rather than accelerating the concentration of property, the tax led to the parcelization of large farms, an increase in the number of property owners, and a decrease in average farm size." However, this does not mean that there were not significant reductions in access to land in the form of usufruct and subsistence capabilities. See Ayala and Bergad, "Rural Puerto Rico in the Early Twentieth Century Reconsidered: Land and Society, 1899-1915," *Latin American Research Review 37*, no. 2 (2002): 65-97.

8. The international organization, La Via Campesina, is credited with coining the term and defines food sovereignty as a system in which local agricultural production is prioritized alongside access of peasants and landless people to land, water, seeds, and credit.

9. This was reiterating the 500 Acre Law embedded in the Foraker Act of 1900, which sought to limit corporations to owning a maximum of five hundred acres but was frequently violated. Some critics of the law suspect that this part of the plan was hatched to prevent Puerto Rico's azucareras from competing with US sugar beet production.

10. The parcelas program ultimately provided some 51,157 plots, averaging 1.04 acres each, to Puerto Ricans. The program was halted in 1945 due to criticisms that the parcelas created "rural slums" and for lack of funding.

11. Ironically, the Populares' slogan would forever be "*Pan, tierra, y libertad*," or "Bread, Land, and Freedom," and an emblem of the *jíbaro* wearing a traditional straw hat.

12. Vivian Carro-Figueroa, "Agricultural Decline and Food Import Dependency in Puerto Rico: A Historical Perspective on the Outcomes of Postwar Farm and Food Policies," *Caribbean Studies 30*, no. 2 (2002): 77-107.

13. This also set the stage for the mass sterilization effort that swept through the Puerto Rican countryside at the hands of local officials swayed by United Statesian rhetoric. See Ana Maria Garcia's *La Operación* (1982).

14. The only company that saw an increase in cane production between 1963-72 was Guánica's South Porto Rico Sugar Company, finally shuttering its doors in 1982.

the shifting United Statesian economic and militaristic demands and pressures that have shaped our modern society.

As a collective, we broke our backs on the colonizers' sugar plantations so they could turn a profit. We suffered from anemia; we worked under scalding sun and through thunderstorms, wet and shivering in the hills of the dewy *cafetales*—we nearly starved. It is one of our many shared tragedies that the lands of the sugarlords lie fallow or were sold off to the multinational semilleras. Even more disparaging is the fact that those corporations occupy the same terrain where the US military first stepped foot on the island of Puerto Rico, and that, 122 years later, our most fertile lands still remain out of our grasp. This land holds our trauma, it has our blood on it, it is our birthright. This land has been calling her long tree frog song, and it has always been time to come home. ○

15. The Agricultural Census fails to capture the multitude of young agroecological farmers, especially women, who are interacting with farming in myriad ways and starting projects—some on leased or family lands—throughout the island, in some cases as a direct response to the food insecurity crisis experienced during and after Hurricane María. Nonetheless, nearby Cuba's proportion of women working in agriculture was roughly 9 percent (of total females in the labor force) in 2011, while Puerto Rico's sat at .4 percent in the same year, according to the International Labour Organization.

16. Only established farmers are eligible for the farm ownership loan via the Farm Security Administration, and only one project has ever been funded in a US territory for *new* farmers under the USDA's Beginning Farmer and Rancher Development Program. Some local agencies (Departamento de Agricultura (DA), Autoridad de Tierras) target larger, conventional farms. DA's Fincas Familiares program once offered an usufruct-to-own structure for farmers to acquire land for the first time. When I spoke to a program administrator, they indicated that all of the available lands were allotted before the program was discontinued several years ago, despite the fact that the DA still owns more than 52,000 cuerdas throughout the island.

17. The cuerda is customarily used in Puerto Rico in land measurements and records, and is equivalent to .971 acres.

18. The term Boricua is used here to refer to Puerto Ricans, Puerto Ricans of mixed descent, and members of the Puerto Rican diaspora.

19. As of 2017, five hundred high-net-worth individuals attracted by the Puerto Rican tax haven laws, Acts 20 and 22, own more land than two million Puerto Ricans combined.

20. Eliván Martínez Mercado, "El boom de Monsanto y las Semilleras Estalla en el sur de Puerto Rico," *Centro De Periodismo Investigativo*, May 11 2017.

21. Nils McCune, Ivette Perfecto, Katia Avilés-Vázquez, Jesús Vázquez-Negrón, and John Vandermeer, "Peasant balances and agroecological scaling in Puerto Rican coffee farming: Crisis, coffee, and agroecological scaling in Puerto Rico," *Agroecology and Sustainable Food Systems* 43, No 7-8 (April 2019): 1–17.

22. Nutrition Assistance for Puerto Rico (NAP), or Programa de Asistencia Nutricional (PAN), commonly known in Puerto Rican Spanish as cupones, is a federal assistance nutritional program provided by the USDA.

23. The Federal Emergency Management Agency of the United States Department of Homeland Security.

24. A Latin American model of bringing multiple farmers together for labor collaboration.

25. Land reform laws were passed between 1959 and 1963 in the wake of the Cuban Revolution. On May 17, 1959, the Agrarian Reform Law drafted by Ernesto "Che" Guevara went into effect, limiting the size of farms to 3,333 acres and real estate to 1,000 acres—any larger holdings were expropriated by the government and redistributed to peasants in 67-acre parcels or held as state-run communes.

26. The reforms passed in 1952 during the Guatemalan Revolution, which redistributed fallow lands of greater than 224 acres to local peasantry, were negated by the US-backed coup in 1954.

27. According to the 1988 Constitution of Brazil, the government is required to "expropriate for the purpose of agrarian reform, rural property that is not performing its social function."

28. The Landless Worker's Movement has coupled land reform with educational reform based on the ideas of Paolo Freire, promoting his concept of conscientizaçao, which highlights the intersectional forces affecting rural communities' condition as historically oppressed peoples.

29. The United States–imposed fiscal control board, or *junta*, as it is referred to locally, took its seat in our capital in 2016 and proceeded to enact a privatization and austerity program that continues to attack public education and healthcare on the island from every angle. The junta has yet to propose any plan for reactivating the agricultural economy.

↑ **Paola de la Calle**   *Just A Banana It Ain't*

News Bureau Offices:
Poston I   56 - 8-A
Poston II   215 - 8-A
Poston III   317 - 8-A

SERIES NO. 117

Managing Editor
For Today
SUSUMU MATSUMOTO

NOW WE HAVE THE
ANSWER...HO HUM

Curious Postonians who were wondering what those pretty little sinks in their lat rines were for, will now have their curiosi ties "checked."

According to informa tion received from U.S. Engineers' office, each men's and women's rest room in Units 1, 2, 3, will be provided with one "slop sink" (so they are called).

The heavily enameled sinks will have both the hot and cold water. It will be used to clean individual bed chambers.

--0--

ZA ZEN TRAINING
(SELF CONTEMPLATION)
IS ON SEPT. 21-26

All young Buddhists of Poston I are re- quested to register for Za Zen training im mediately, to be held between Sept. 21 and 26, from 5 to 6 a.m. at Blk. 45-14B.

Za Zen (self contem- plation) training is "for yourself and others in enlightening a better and happier daily living," accord- ing to the Buddhist teachings.

EDITORIAL

VOICE OF AN ISSEI

The writer has made a somewhat bold remark in this column last Wednesday in respect to issei's ability in both business and farming undertakings and said that they have been the producing factors until they came here for the duration. Actually, our older farmers have made the entire State of California productive and kept it green all year around.

The set-up of farming project here, we learn ed, is after a very good pattern--rather an i- deal one. However, it lacks encouragement. Co- operative system applied to it is quite a new one, and no farmer can swallow it especia lly when they get paid in the lowest bracket of monthly allowances of measly twelve dollars. The farmers should be placed above all othe rs in regard to the amount of allowances. the wages or allowances should be so made accord- ing to the degree of importance the job implies.

Enough of this. But here is a tentative plan which if approved, might eliminate unpleasant - ness and dissatisfaction now prevailing among our farm hands and give them something to look forward and at the same time, speed up the pro- duction.

Here it is.

1. The farming project in Poston should be turned over to and under the direct super vision of the Federal Department of Agri- culture to facilitate the general market- ing and distribution of products.

2. Leasing of tract of land to any farmer for the period of five years on share crop basis. The Government, in leasing the tract of land on share proposition, agrees to furnish and loan all the nec- essary farming implements, if asked and also furnish seeds, fertilizer and insec- ticide, free of charge.

3. The Government reserves the right to choose the kind of crops wanted.

4. All the farm hands hired by the individual farmers should be paid once a month by the Government said allowance or wages should be the highest available for en- listee or worker in the Center.

5. That the Government, as a trustees, handle all the products keep the account of market quotations from time to time on all shipments, chages and other item s and make a yearly report and pay the bal- ance in cash to individual farmers for their dues. The amount of wages payable

Cont. or page 3.

The *Poston Chronicle* was the main newspaper of the Poston, Arizona concentration camp, where thousands of Japanese-Americans, mostly from southern California, were incarcerated during World War II. The largest of ten concentration camps operated by the War Relocation Authority, Poston was built on the Colorado River Indian Reservation over the objections of the Tribal Council. It had an expansive farming program, to which Kuni Takahashi suggests several improvements—better wages!—in his opinion column, "Voice of an Issei." The *Chronicle* also reported on inmates leaving for work, including to assist sugar beet farmers outside the camps. The paper's final issue was released on October 23, 1945.

Jamie Hunyor

A FARMHAND'S PERSPECTIVE

I work on an organic vegetable farm tucked into the foothills of the Sierra Nevada. The 2020 growing season will be my third. The 5.5-acre operation on Nisenan land is situated on the last 150-acre parcel of a historic ranch. The farm has a 100+-member CSA, wholesale accounts, and a farm stand.

Working in small-scale agriculture can be challenging because both its pros and cons exist in a space that lacks large financial gains. The benefits, in my mind, exist entirely outside of the capitalist framework.

I have no idea how to quantify right livelihood, proximity to nature's beauty, and the slow pace of seasonal, rural living. The downsides are much easier to count—long hours, low wages, no healthcare—and, due to our socialization in a capitalist society, these are easy to fixate upon.

Within the framework of capitalism—the system—farmers work toward change from inside the belly of the beast. I have friends who work on farms in California, the Pacific Northwest, the Midwest, New England, and Appalachia. Many of us were turned on to small-scale agriculture in the college town of Athens, Ohio and then spread out across the country. Almost all of our passionate conversations circle back from heady paradises of utopian ideals to a common refrain: our pay.

Not one of my farmhand friends is offered healthcare. A handful of them pay higher taxes on their annual income because their bosses classify them as independent contractors to save money on their own taxes. Some

of us pay part of our income back to the very people who write us our checks so we can live in rustic structures on their land. By necessity, we become skilled at squirreling savings away, building up a stash of acorns from our small paychecks.

I am fortunate to fall back on subsidized healthcare from the state of California, but recognize not everyone has this option. I know of one farm in a thirty-mile radius that offers health insurance and 401(k) benefits to its employees. There's even a farm in Placerville that operates as a cooperative—a worker-owned farm that offers employees the opportunity to buy in after a year of employment. Some of us on the smaller farms talk of banding together to coordinate purchasing a group plan, but that dream slips through the cracks without someone stepping up to organize it.

The problem small-scale agriculture is butting up against is the question of value. What is an hour of labor worth? What is a head of lettuce worth? What is the worth of choosing to grow crops in ways that don't deplete the soil but add to its health instead? What is the worth of biodiversity on the land that is being farmed? What is the worth of a workday that the worker enjoys thanks to engagement and collaboration instead of dreading due to menial repetition? Is it worth it to a community to pay a higher price for produce in order to raise not only the farmer's standard of living, but also their employees'?

Each small farm is its own unique system, a microcosm working both with and against the

dominant culture's expectations of how a business should be run, and each will answer each of these questions differently based on its own context. That's where the beauty of this work lies: rather than regurgitating what capitalism has offered us through an extractive and destructive food system, we are offered the opportunity to create systems that are specific to the land we work, our life experiences and those of the people we work with, and the desires of our unique community of customers.

Online, I often come across critiques of the "grind culture" that our society has normalized via capitalism, and to consistently have fun at work seems to me a direct challenge to the idea of grinding. Consider the language: grinding is an action of violence, crushing or reducing something from a whole into powder or parts. I don't see any laughter there, and while laughter doesn't pay the bills, I'd rather make do and enjoy than make a killing indoors doing work I'm not wholly engaged in.

This is my anti-capitalist practice. I have the privilege of preserving my own food from the farm's abundance. I have the opportunity to engage in a small gift economy by exchanging homemade goods and skill-sharing with my coworkers. As Wendell Berry did, and countless women before and after him, I believe that choosing to stay home (or close to it) is inherently anti-capitalist, as it encourages autonomy through domestic practices like cooking and entertaining oneself.

I live in a yurt on the farm near three gorgeous rivers that I swim as often as I can in the summer months. I have access to all the fresh vegetables I could ever eat, as well as an abundance of right livelihood. Working alongside my bosses and coworkers each day makes the work like play. ○

Jon Henry

EGGONOMICS

To help pay my way through two art grad programs, I became a produce huckster in the summers. I attended produce and plant auctions, farmers markets, harvest festivals, and ran a roadside stand, which I still keep going. I moved between the Anabaptist produce and high-end art realms of the Mid-Atlantic.

Photos by **Corinne Diop**

I began to wonder how the art auctions at Christie's differed from the produce auctions in the Shenandoah Valley in terms of value, process, honesty, manipulation, currency, supply chains, or impact on ecosystems.

For an installation at the Arts Council of the Valley, I tapped into my ongoing quantification of food through the complicated laying of labels. As an artist, I was fixated on the process for certification as "Predator Friendly," "Carbon Neutral," and "Pasture Raised." I wondered who went to the farms to ensure that the farmer wasn't killing the possums that had killed their chickens, or was letting the chickens move freely over sprawling green acres. As a huckster, I was curious about the value of such labels and what was needed to create value for my markets: cage free vs. free range or organic vs. local farm. On both fronts, I remain curious on pricing to sell: How can one decide the exact or true value of an egg, painting, burger patty, second edition print, or tomato slice?

Labels create value hierarchies—they may have similar meanings, but reflect totally different levels of invested work, resulting in different price points. At the farmstand, I found the sweet spot that met my customers' minimum demands for free-range eggs while still getting me a return.

For the installation, I worked with a friend to source locally pastured, organic, soy-free, non-GMO, predatory friendly, humanely raised, carbon-neutral eggs in compostable and recycled cartons. The resulting product featured my signature atop every egg and was priced at

$13 a dozen. We sold a few eggs but never sold out of the eggs, unlike at my farmstand, where I courted the regular, middle-income customer rather than a wealthy clientele.

I ate both the $13 dozen and my farmstand's $2.50 eggs and couldn't exactly tell the difference except by the yoke's color.

I know my regular customers would revolt if I tried to up egg prices drastically; I already survived a pushback against a 25-cent raise in price. Even my roadside farmstand in the suburbs of DC couldn't handle regular retail prices above $3. One Northern Virginia competitor (who has since closed) offered a DIY Egg Dozen: you chose the eggs for your dozen, paying $8

for that freedom. We accept donated cartons at our stand, and one day a prestigious client dropped off a carton from that competitor. I jokingly asked why they had been "cheating" on me, and he retorted that it was a lesson learned—his family couldn't taste the difference. It was just a cute experience to get some blue eggs mixed in with their regular brown eggs.

As a white rural person living below the Mason-Dixon Line, I remain skeptical of overly labeled products and instead look for the understated items. It might be the years of reading post-structuralism, but at some point an egg just needs to be scrambled. ○

Elizabeth Henderson

ROOT SOLUTIONS TO CRISIS FOR FAMILY-SCALE FARMS

The finally inescapable standards by which agricultural choices must be made are the ecological health of the farm and the economic health of the farmer.

—Wendell Berry

I started farming in 1980. For as long as I can remember, most of the farmers I know have supported their farms through off-farm work that provides health insurance and other benefits: teaching, or work in healthcare. It's a condition of an ongoing farming crisis in the United States: under relentless and steadily increasing financial pressures, dairy farmers sell their cows and turn to field crops, raising cattle for beef, or selling hay—anything to keep the farm alive. Talented young farmers give it their all for five, even ten, years before quitting. Experienced farmers (including organic farmers who receive slightly higher prices than their conventional neighbors) go out of business, sell what they can, and find "real" jobs. Developers gobble up farmland, which has grown too expensive to buy with farm earnings. The price farmers receive for crops does not cover the costs of keeping farms viable, much less the extra costs of developing and sustaining ecological or regenerative systems.

The 2018 Farm Bill[1] barely touched the structural and fairness issues that led to this ongoing disaster for family-scale farms and the food security of this country. These pervasive issues—a farming crisis—have only been highlighted by the COVID-19 pandemic.

Despite a shortage of farmworkers, wages remain below the poverty line. People of color and women are still trapped in the lowest-paying food system jobs, many forced to survive on Supplemental Nutrition Assistance Program payments. Being recognized as "essential" to the pandemic economy has not yet resulted in higher pay or better working conditions. The president's tariff game has only made things worse: billions paid out to compensate for trade losses made big farms even bigger, with grave consequences for the environment.

A Historic Solution:
Parity Plus Supply Management

Wendell Berry offers a possible path, saying:
  The problem that has impoverished and destroyed farmers nearly always is that of low

Dorothea Lange *Migrant agricultural worker's family. Seven hungry children. Mother aged thirty-two. Father is native Californian. Nipomo, California, March 1936.*

↑ **Vera Bock** *Work Pays America! Prosperity.*

prices resulting from surplus production. That is also, obviously, a land-destroying problem. The only solution to that problem that can sustain the small farmers is the combination of production control and price supports as exemplified by the Burley Tobacco Growers Cooperative Association as it was reorganized in my region under the New Deal in 1941.[2]

What does "production control and price supports" mean, and how did they work under the New Deal?

In the depths of the Great Depression, so many family farms were going bankrupt that the federal government stepped in to help them avoid eviction and to increase crop prices. The Agricultural Adjustment Act (AAA) of 1933 declared an economic emergency "being in part the consequence of a severe and increasing disparity between the prices of agricultural and other commodities," justifying action as being in "the national public interest."[3]

The AAA established the parity system of pricing and supply management to reestablish farmers' purchasing power, using the years just before World War I—during which balance existed between farm earnings and the prices farmers had to pay for inputs and equipment— as a base period. Retail prices to consumers were also pegged at the same proportion of consumer income as during those pre-war years. To raise prices for farm products, the AAA reduced oversupply by establishing marketing quotas on the acreage farmers could use for basic commodities: wheat, cotton, field corn, hogs, rice, tobacco, rye, flax, barley, grain sorghums, cattle, peanuts, sugar beets, sugarcane, and potatoes. Conservation practices were required on the land that was taken out of production. That first year, some crops were even plowed under. There were also marketing agreements that controlled the quantity, quality, and rate of shipment to market, effectively limiting production of some fruits and vegetables. The Secretary of Agriculture was also enjoined to let the president know if imports threatened to reduce prices to US farmers.

Due in part to these programs, farm income in 1935 was more than 50 percent higher than in 1932.[4] Although agribusiness successfully brought suit against the first version of this parity system, the revised approach set up by The Soil Conservation and Domestic Allotment Act in 1936 proved more durable and lasted through the 1960s.

Farmers were free to participate or not, and could vote down marketing quotas. County committees were established to provide a forum for local referendums by commodity. These still exist and hold elections every year among farmers who participate in government programs. In some parts of the country, these committees work well; in others, the committees have made racist decisions, such as providing operating loans to white farmers in the spring when farms need start-up money, but not paying black farmers until fall. Some committees have discriminated against farmers who use organic methods, refusing disaster payments because the organic farmers did not use chemical protectants for crops.

In "Crisis by Design: A Brief Review of US Farm Policy," Mark Ritchie and Kevin Ristau[5] summarize the three central figures of the parity system:

It established the Commodity Credit Corporation (CCC), which made loans to farmers whenever prices offered by the food processors or grain corporations fell below the cost of production. This allowed farmers to hold their crops off the market, eventually forcing prices back up. Once prices returned to fair levels, farmers sold their crops and repaid the

CCC with interest. By allowing farmers to control their marketing, the CCC loan program made it possible for them to receive a fair price from the marketplace without relying on subsidies.

1. It regulated farm production in order to balance supply with demand, thereby preventing surpluses.
2. It created a national grain reserve to prevent consumer prices from skyrocketing in times of drought or other natural disasters. When prices rose above a predetermined level, grain was released from government reserves onto the market, driving prices back down to normal levels.

From 1933 to 1953, this parity legislation remained in effect and was extremely successful. Farmers received fair prices for their crops, production was controlled to prevent costly surpluses, and consumer prices remained low and stable. At the same time, the number of new farmers increased, soil and water conservation practices expanded dramatically (eliminating the Dust Bowl), and farm debt declined. The program was not a burden to the taxpayers, either: The CCC charged interest on its storable commodity loans and made nearly $13 million in two decades."[6]

Through four decades of changing conditions—depression, war, post-war prosperity—the combination of parity with supply management and required conservation practices worked for farms and rural communities.

Corporate Dominance Ended Parity
From the end of World War II through 1974, a consortium of agribusiness, banking, and university leaders deliberately set out to eliminate parity with policies that cut farm prices to drive excess "resources"—meaning, farmers and their families—out of the countryside.[7] By the

mid-1970s, farm prices were dropping and farm numbers decreased rapidly.

Farm earnings today would be very different if parity pricing levels were still in place. According to the National Agricultural Statistics Service, the parity price for one hundred pounds of milk in May 2019 would be $52.80, and a bushel of corn would be $13.20. Instead, conventional farmers were getting $18 for a hundredweight of milk and $3.63 for a bushel of corn.[8] With its combination of subsidy and emergency payments to commodity farmers along with crop insurance, the 2018 Farm Bill enshrines cheap food policy with low farm prices that mainly benefit the biggest agricultural corporations.[9] Until the early 1970s, those corporations had to pay farmers decent prices in the marketplace. Since then, it has been the taxpayer who covers the costs of cheap food while the corporate buyers who purchase most of these crops make out like bandits. This adds up to a major transfer of wealth from the farmers and the public to the likes of Amazon, Walmart, Tyson, and Archer Daniels Midland.

What a Green New Deal (GND) Could Do for Agriculture
Farming organizations including the National Family Farm Coalition (NFFC) and the National Farmers Union have continued to demand a return to parity and supply management. But for twenty years or more, this set of policies has been deemed unlikely to gain traction among lawmakers in Washington, DC. Then, in a flash of light, the 2019 Green New Deal resolution by Senator Edward J. Markey and Representative Alexandria Ocasio-Cortez made it realistic once again to consider this set of root solutions to the food and farm crisis.[10] Their resolution, however, is just an outline calling for a select committee that will be tasked to write the actual

United Farm Workers boycott lettuce, 1965–1980

legislation. It is up to us as organic farmers and food activists to put together concrete proposals based on our lived experience.

Iowa farmer and NFFC steering committee member George Naylor explains the inherent logic of parity plus supply management:

> When a farmer is given a quota that sets the limit of whichever storable commodity can be marketed or fed to livestock on the farm along with a parity price, the incentive to produce as much as possible with whatever technological inputs and neglect of the land disappears. The new logic would be to produce only the quota, and spend as little as possible on inputs, and engage in as much conservation as possible.[11]

The higher price on a set amount of production stops farmers from overproducing and provides an economic incentive to use the most ecological and efficient practices.

While we can learn a lot from the parity sys-

tem of the old New Deal, both its strengths and also its failures (especially in regard to farmers of color), a new version is needed that meets the conditions of the 21st century. This Green New Deal for Farmers must include racial justice and equity in the safety net it provides for farms. Family-scale farmers, regardless of color, need a system of fair pricing—that is, prices that cover the real costs of living and farming, including conservation practices that regenerate natural resources. I can imagine an exciting public process where groups of stakeholders all over the country hammer out the details. On a much smaller scale, that is what we did in the 1990s to launch the National Campaign for Sustainable Agriculture. I helped Alison Clark, founder of the New York Sustainable Agriculture Working Group, organize five regional hearings around the state, where a few hundred farmers and activists brainstormed and formulated recommendations that were then combined with similar reports from meetings in other states.

Here is a rough first draft:

Twenty-first century parity will cover the basic commodities and reestablish farmer-held reserves for grains as buffer stocks that protect farmers from poor harvests, climate disasters, or price volatility.

Our rural Green New Deal will increase incomes for all food and farmworkers. Growing food justly and sustainably is expensive. Instead of driving down the costs of farming to make food cheap enough for urban workers to buy on stagnating wages, all workers must make enough to afford sustainably produced food. Consumers must be able to pay for the knowledge embedded in, and carbon sequestered through, low-input, sophisticated agroecological farming using renewable energy. And farmers and farmworkers must be paid fairly and appreciated for their work.

Parity pricing and supply management should also be extended to fruit and vegetables, "speciality crops" in Farm Bill language. Since fruit and vegetables are perishable, the GND will invest in value-added enterprises that could be farmer- or worker-owned cooperatives in every county where these crops are grown. If excess supply threatens to lower prices, the fruit and vegetables would be frozen, canned or dried, or made into products that can be stored for use year-round.

Investing in local and regional processing would stimulate local economies and provide many jobs. The GND would return livestock to family farms, reversing the shift to large-scale Confined Animal Feeding Operations (CAFOs) that have eliminated the need for diverse crop rotations. Family farm livestock production integrates crops and livestock for a much more flexible and resilient system that reduces the pressure for routine antibiotic use. This system also increases biodiversity and strengthens a farm's economic viability by adding opportunities for new workers while improving the quality of the meat, milk, and eggs.

Farmers need contract reform. Those who sell to bigger entities need legislation to protect their rights to freedom of association so that they can form groups or cooperatives to strengthen their bargaining position in negotiating fair contracts without threat of retaliation. In addition, a limit must be set on the middlemen's share of the final shopper dollar. If prices go up, middlemen must pay farmers more; if the prices processors pay to farmers go down, the point-of-purchase price for shoppers should follow. With control by mega-corporations an ever greater threat to family-scale farming, the GND must be linked with anti-trust measures like the Booker bill that calls for a moratorium on mergers.[12]

All farmers should be eligible for GND programs whether they own land or rent it with cash payments or through sharecropping, and our solution should include measures that are essential to establishing farm work as a respected and fairly remunerated profession. Ocasio-Cortez wants to guarantee living wages and green jobs—that guarantee must cover the jobs on farms. Since farmworker advocates and Department of Labor staff have found that over 50 percent of farmworkers on US crop farms are undocumented, immigration reform based on human rights needs to accompany the GND. Farmworkers should have the option of a path to citizenship if they want to live in the US and the freedom to come and go across the border to visit their families back home.

Like farmers, farmworkers need freedom of association so that they can form groups or unions to negotiate fair pay and working conditions. If farms are guaranteed prices that cover their costs of production, farm earnings will be high enough to pay farmworkers time-and-a-half for overtime over forty hours a week, like workers in almost every other sector.

This first draft includes ideas drawn from many places: from listening to my fellow organic farmers, farmworkers, and to other food chain workers, and from reading agricultural history, and by keeping up with the news. But this cannot replace a noisy, joyous, contentious democratic process where working people hammer out solutions together.

In "The Green New Deal: Fulcrum for the farm and food justice movement?" Eric Holt-Gimenez writes:

Social movements have an opportunity to join together as never before—not just to get

behind the Green New Deal—but to form a broad-based, multi-racial, working class movement to build political power. Visionary leaders from these movements are already knitting together strategies for solidarity, education and action.... The Green New Deal just might be the fulcrum upon which the farm, food and climate movements can pivot our society towards the just transition we all urgently need and desire.[13]

That is the challenge we face: to pull together a big enough movement of farmers, farmworkers, labor unions, Indigenous peoples, environmentalists, faith communities, youth, and rural and urban activists of all kinds to transform this climate emergency—linked so closely to the coronavirus pandemic—into an all-out campaign to save human life on this planet and create an ecological and just civilization. ○

Notes

1. The Farm Bill, or Agriculture Improvement Act, is the compendious package of food and farm policies that is renewed every five years and governs the programs of the US Department of Agriculture, including the nutrition programs that account for 80 percent of the funding.

2. Gracy Olmstead, "Wendell Berry's Right Kind of Farming," *The New York Times*, October 1, 2018.

3. Wayne D. Rasmussen, Gladys L. Baker, and James S. Ward, "A Short History Of Agricultural Adjustment, 1933-75," Economic Research Service USDA, *Agriculture Information Bulletin, no. 391* (March 1976): 5.

4. Rasmussen et al, 4–6.

5. Mark Ritchie and Kevin Ristau, "Crisis By Design: A Brief Review Of US Farm Policy," League of Rural Voters Education Project (1987): 2–3. Also see Patti Edwardson Naylor, George Naylor, and Ahna Kruzic, "Parity and Farm Justice:

Recipe for a Resilient Food System," *Food First Backgrounder* 24, no. 2, Summer 2018.

6. The US parity system resembled existing dairy supply management, which came under attack by President Trump during the 2018–19 NAFTA negotiations. For a description of the Canadian system, see Hannah Monicken, "U.S. singles out supply management in Canada's WTO trade review," *Coalition for a Prosperous America*, June 13, 2019.

7. Mark Ritchie, "The Loss of Our Family Farms: Inevitable Results or Conscious Policies?", League of Rural Voters (1979).

8. USDA National Agricultural Statistics Service, *Agricultural Prices ISSN: 1937-4216*, June 27, 2019.

9. From Iowa farmer Brad Wilson email to Comfood listserv of June 26, 2019: "... farm programs 'PIT CHEAP PRICES TO BENEFIT AGRIBUSINESS AGAINST CONSERVATION'... The major thing the farm programs do today...is that they

allow chronic free market failure. And secondarily, they cover it up with inadequate subsidies that would otherwise not be needed at all..."

10. Raj Patel and Jim Goodman, "A Green New Deal for Agriculture," *Jacobin* (April 4, 2019).

11. George Naylor, email to author, July 26, 2019. Naylor writes about parity at greater length in "Without Clarity on Parity All you Get is Charity," *Food Movements Unite!* (Oakland, California: Food First Books, 2011).

12. US Congress, Senate, *Food and Agribusiness Merger Moratorium and Antitrust Review Act of 2018*, S.3404, 115th Congress, introduced in Senate August 28, 2018.

13. Eric Holt-Gimenez, "The Green New Deal: Fulcrum for the farm and food justice movement?", *Food First*, December 15, 2018.

The Twelve Principles of EPIC

(End Poverty In California)

1. God created the natural wealth of the earth for the use of all men, not of a few.

2. God created men to seek their own welfare, not that of masters.

3. Private ownership of tools, a basis of freedom when tools are simple, becomes a basis of enslavement when tools are complex.

4. Autocracy in industry cannot exist alongside democracy in government.

5. When some men live without working, other men are working without living.

6. The existence of luxury in the presence of poverty and destitution is contrary to good morals and sound public policy.

7. The present depression is one of abundance, not of scarcity.

8. The cause of the trouble is that a small class has the wealth, while the rest have the debts.

9. It is contrary to common sense that men should starve because they have raised too much food.

10. The destruction of food or other wealth, or the limitation of production, is economic insanity.

11. The remedy is to give the workers access to the means of production, and let them produce for themselves, not for others.

12. This change can be brought about by action of a majority of the people, and that is the American way.

"I say, positively and without qualification, we can end poverty in California," wrote Upton Sinclair in 1933. "I know exactly how to do it, and if you elect me Governor, with a Legislature to support me, I will put the job through—and I won't take more than one or two of my four years." Sinclair ultimately lost one of the most heated elections in the state's history, but some of his ideas lived on, if less radically, in the federal work relief programs of FDR's New Deal. And his vision of state-managed cooperative colonies, based on an economy of "production for use" instead of "production for profit," found some expression in the Farm Security Administration's 1930s experiments with rural co-ops and collective farms. Learn more at the University of Washington's online archive of the EPIC campaign.

Kelly Garrett

SLUGS

MAYBE WE ARE ALL INSIDE A SLUG,
LIKE HOW THE WORLD IS IN A MARBLE
AT THE END OF *MEN IN BLACK*.
MAYBE THE SLUG MOVES SLOW ON PURPOSE
TO KEEP US UPRIGHT,
AND IF SALT IS POURED,
WE SWEAT AND CRY OUT
FOR AIR, FOR WATER, FOR SHADE.
AND IF THERE'S AN EARTHQUAKE,
A CHILD MAY HAVE TOOK US FOR A WALK,
AND MAYBE THE SALT THAT FALLS OFF OUR BODIES
MELTS ICEBERGS, AND THAT EXTRA WATER
MAKES THE SLUG BLOAT—PAST ITS CAPACITY.
TOO MUCH PRESSURE HAPPENS.
WE DON'T KNOW WHERE TO START.
THE SLUG COULD BURST
WITH ANY WRONG MOVE.

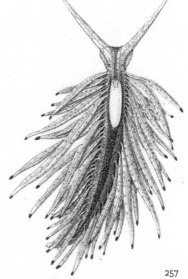

Emily Ryan Alford

GROWING FOOD ON VENICE BOULEVARD

An Argument for Fenceless Gardens and Public Commons

I was on my way to meet a reporter at the garden when I saw the big, white dump trucks. Police officers and folks in white hazmat suits crowded the small encampment down the block. *Fuck fuck fuck fuck fuck fuck*. A sweep.

I toured the local journalist around our third-of-an-acre slice of public, Los Angeles city property—occupied Tongva land—and spoke about our urban gardening internship for youth experiencing homelessness. As a service provider and food grower, I lead this paid workforce development program through the lenses of justice, equity, and healing. Together, we grow food, learn how to cook it, and feed others. Interns receive marketable job skills and support transitioning into long-term employment. As well, we witness a strong, positive correlation between gardening and mental health in youth participants. Garden interns frequently report reduced anxiety and an increased sense of confidence, groundedness, and calm—an experience I've had in my own body, and a large reason I come back to farming year after year.

Just the week prior, I told the journalist, I'd walked through the neighboring encampment now swarming with sanitation workers and police, and noticed, outside of a resident's tent, a vase of flowers harvested from the garden.

Our garden's fencelessness makes the space and all we grow available to everyone.

The journalist left and I began a batch of compost tea. A man walked into the garden, someone I'd seen before but didn't yet have a rapport with. Distressed, he told me that they'd taken all of his clothes and belongings. He'd just done laundry. They took it all but left behind trash. I listened, offered water.

He was walking away when he said, "I'm sorry, I harvested all of your Sugar Baby watermelons this summer. They were juicy and delicious, and fed a bunch of us."

My face brightened. I told him how glad I was that he'd enjoyed them and assured him that an apology was not necessary. I asked his name, told him mine.

Then he said, "I just need to cry."

He stood next to the sunflowers; tears fell. I walked toward him and asked if I could put my hand on his shoulder. We stood like that for moments. Cars continued to buzz down Venice Boulevard.

I asked him what he thought of the budding yellow blooms, and if he'd like to see our purple cauliflower. We walked through the garden, I harvested a fragrant marigold. When I handed the orange pom-pom to him, he again began to cry. He asked if I knew that you could eat

Nikki Mokrzycki *Edibles*

dandelion greens, I told him I loved them. We sat down on the earth and ate the spiky leaves together.

I cried after he left, as I do often in this work. I cry as a witness to LA's housing crisis, which has left over 60,000 people experiencing homelessness. I hope I never cease to be moved by human connection, nor grave, systemic injustice.

Public commons offer gifts of connection, health, and healing to our communities, especially when we prioritize the safety and well-being of folks who are marginalized. I plant more seeds—sunflowers, bright tall

shining branching golden. When a cabbage disappears overnight, I trust that it's been enjoyed by an unhoused neighbor. I say hello to everyone who walks by. I dream of more fenceless, green public commons in urban centers, and urge my community to do the same. Butterflies and humans and hummingbirds—all of us— need green space, need shade, need fresh food, need connection, need gathering places, need to invest in healthy soil and the beings who live in and from it.

Resist enclosures. Grow plants. Build relationships. Reclaim public commons. ○

↑ *Don't Buy Bombs When You Buy Bread!* between 1965 and 1975
→ *Outdoor amphitheater in Norris Park*, Tennessee, between 1933 and 1945

(Built) Landscapes

Transportation Departments

Nina Pick

COMING DOWN
FROM THE
TOWER OF BABEL

When they had built the city
and the tower in it
that stretched all the way
to the heavens
they forgot what it felt like
to put their feet in the dirt.
Their hands, rough from stacking bricks,
had forgotten the softness of a patch of grass,
the innumerable textures of spine, spindle,
nettle, fur, the brusque nudges
of sheep, the gentle, toothless nibbling
of newborn goats.

So long ago had they built their tower
they could no longer remember
the earth that their bricks were made of—
even the sensation of wet clay
between fingers, between toes,
they had forgotten.
But their bodies knew, could feel
the continuous song of river and rain,
could feel the ancestors beating the bone-drums,
could feel the ewes, the rams, on the hillside singing.
The God-in-the-body knew.

And the God-in-the-body rebelled against
forgetting,
rebelled against brick,
rebelled against mortar,
and taking the people from their tower
scattered them
like a basket of seeds,
like a handful of wildflowers,
over the face of the Earth,
until at last, lowering their faces to hers,
they remembered.

Christie Green

$1.67

I sit at the oval-shaped conference room table with the men: a developer, an architect, an engineer, and a contractor. Some call me a landscaper, others call me a landscape designer, or landscape architect—a title I'm allowed to claim now that I have an advanced degree, have passed four national exams that cost between three-hundred-fifty and five-hundred-fifty dollars each, and satisfied licensure application requirements. This title validates me as someone who designs for "the health and safety" of the public. I can charge a higher hourly rate and put three letters behind my name: PLA, or Professional Landscape Architect.

"I can't stand edibles, Christie," I hear from the developer. "And fruit trees. I HATE fruit trees. They're so messy. You got birds eating them, pooping everywhere, messing up cars, fruit falling on sidewalks. Edibles are just a nice idea." He brought me on to help the architectural design team think "big," think "outside the box," to "wow 'em" in the design competition.

"You've got $1.67 per square foot to work with, Christie," the contractor states flatly, eyes cast down, scanning the spreadsheet in front of him. He's flown in from a nearby booming Southwest town with his pressed white shirt, navy slacks, and practical loafers. He's very white with slicked-back, strawberry-blonde hair and a gold wedding band squeezing his ring finger. I imagine his life as pudgy.

I lean back in my chair, arch my back, interlock my fingers behind my head, and drift off to my happy place, the seven-acre farm up north where Olivia was conceived in spring of 2004, when the thirty-foot cherry trees blossomed and burst bushels, almost too many to dry, can, freeze, and bake into pies. I indulge in the memory of the farm whose land stretched to the Rio Grande, where the herds of elk grazed in December morning fog, ephemeral ghosts landing on lush alfalfa after having crossed Black Mesa near the Chama–Rio Grande confluence.

On the farm, I allowed the plum thickets to cover the acequia bank, barely pruned the yearling heirloom fruit trees, let the orchard grass seed out. While clients in town demanded native penstemon be yanked from the root because they're "so ugly" when dormant, I relaxed into a less manicured rural life. The chickens roamed freely with access to the porch; poop plops on the concrete disgusted my mom. "Why the hell don't you fence that part off, hon? Put those damn chickens in their own yard. They don't need to be comin' into yours!"

My evening walks down the fields through the bosque and onto the shore of the Rio Grande loosened the work day's rigidity. Client demands slipped from my professional, solve-it, spiff-it-up, make-it-pretty-now shoulders. "When will those plants start to grow?" "Won't those be blooming soon? We leave for our California home at noon on Monday. I want to see something happen before we go." I'm supposed to know how to turn nature into a machine. I'm supposed to work her harder, whip her into shape, force her to grow faster so that I can earn more, be worth more, show how much I can do, too.

But here along the river, I can meander with the current, laze in the cottonwood shade, hang chile ristras to dry from the portal on all sides of the house. I grind and package their hollow red shells, taste their spicy, earthen flavor throughout the year in posole, beans, carne adovada, and enchiladas.

These homegrown chiles aren't just for decoration.

The developer chimes in about wanting trees, shade, color pop—something eye-catching. "We need to make this place different, attractive, up-and-coming," he says. I wonder how to even get a shovel on the site for $1.67 per square foot.

"How about fake plants with a remote control?" I smirk.

"Yeah, you got it."

The developer lights up as I adjust the wide leather belt around my waist, push my hair back, bite down on the pencil. My waist thickening with age, and my back getting stronger, some may say I'm pretty; I think I'm not bad looking but have to work hard at whatever pretty I have. I see the lines on my face deepen, age and sun

spots become more prominent, a bit of droop to my neck and chin. You may say I have the beauty of an antique leather bag, not a Chanel that "pops" with gold buckles and shiny patent straps.

The architect and the engineer hurry to the next subject: should the bathroom closets have doors?

We continue to examine the lines on paper, adjusting wall edges and width of sidewalks to squeeze a little more profitable square footage onto the lot. "If we cut that big 'ole cottonwood tree down, we can get at least two more units in," the contractor says.

"Yeah, and that neighbor to the east may sell us his lot," the developer adds. "Nothing there but a bunch of old junk cars. Bet he'd go for a nice check."

I imagine how much food could grow on this urban lot, a stone's throw from the Rio Grande, in the traditionally agricultural South Valley of Albuquerque. I imagine drawing lines on the paper that represent rows of annual crops— corn, beans, and squash, the Three Sisters rooted in generations of rich riparian soil. I see laden fruit trees and roadside neighborhood stands like up north near El Guique, where farmers sell seasonal produce along the highway to tourists and locals hungry for homegrown flavor. What is a bushel basket of food worth compared to a two-story housing development? Does a patch of lettuce, tomatoes, herbs, and onions bring the neighbors and families together differently than the newly planned concrete patio? How can the old ways be worthy if the bottom line sees more profit in "high density" housing?

The developer insists the complex should offer childcare, something for working parents that's convenient. "We could even have a playground centered in front of the apartments, near the laundry mat, so moms can see their kids while folding clothes on the weekend. You know, a Tot Lot. Can you draw that up, Christie? Remember the playground-certified mulch that has to go in the fall zone around the play equipment. We don't want the kids getting hurt. And there will be a six-foot tall fence so they can't get lost down along the river."

While I note the added scope of service to my to-do list, I see in my mind's eye a photograph of me with Olivia from twelve years ago. In the image, I work, shovel in hand, compost in wheelbarrow, tools and truck in the client's driveway. Olivia is about six months old with her sun bonnet and chubby legs poking out from the front pack harnessed to me. This was business as usual: another day at work with Mom.

The client had pulled me aside that afternoon, asking me to put the shovel down, to come to the porch. I was sure he was angry about paying me for time billed to him when I was multitasking as a breastfeeding mom, doubling as a mother and professional.

"Christie, I have to tell you something," he said. For once, his clenched jaws released. "I'm forty-seven years old and am still in therapy to this day because my mom never had me with her, never spent time with me. I was raised by a nanny."

My daughter was born on January 7, winter in northern New Mexico and a time of frozen cash flow for me and any self-employed landscape design-build professional. Two wintry weeks later, I bundled Olivia up in the front pack to take her to work pruning trees. Her little mouth puckered to my nipple while I selected branches to cut, buds to head off. In March, I breastfed her while supervising the guys as they constructed a stone wall and pulled over on the side of the road to change her diaper in the truck, careful to tuck the edges of the soiled wad back into its tape so turmeric-colored baby poop didn't ooze onto the floorboards or seats.

Nearly a year later she was with me again, at one of my first big installation projects up north on a rural property recently purchased by a wealthy couple from back East. It was December, cold, and the time of year, once again, when my bank account slows to a

← Original illustration by **Katy Harris**

dormant crawl in sync with plant rhythms. Work screeched to a virtual halt. I went to check on the final touches of the installation and to get paid. This client, like some others, needed more nudging and reminding that his account was past due.

As he, his wife, and I stood there gazing out to the pasture and the newly installed landscape, admiring his accomplishments as a self-proclaimed "country squire," I swayed my hips side to side, the ingrained motion of all moms with babies, trying to soothe an impatient Olivia, and reminded him of the payment owed. He and his wife looked out to the land as he pulled a check from his shirt pocket and dropped it on the ground in front of me.

"I hate to see a woman beg," he declared.

Bending down with Olivia hugged to my torso, I reached for the check, bit my lip, and thanked him for the payment. The Squire, his wife, and I didn't make room for more words or pleasantries.

I buckled Olivia into her car seat, her blue saucer eyes open to mine, her trust and need softening the bitter brewing in me.

Since then, through fifteen years of land-based work, countless projects, and a wide range of client relationships, Olivia still mends what's been broken. She reminds me of the simple, daily yes: make life through seed, soil, water, blood, bone, root.

We begin to wrap up the meeting, the engineer, architect, developer, contractor, and I. "Let's revise the plans with the new site layout, adding a couple of units to the back corner. Christie, go ahead with the smaller apartment patios, let's get that playground in there. And I want to see a lot of street-front parking and shops. We need to get people in there spending money," the developer says, cheering us on.

He pulls me aside in the parking lot. "You know, Christie, we'll get you a better design fee once we've earned the tax credits and can move beyond schematic design. For now we have to do a kick-ass concept to win the design judges' interest. If we get them turned on to the project, they'll fund the next round of plans and implementation. Then we can pay you more," he says.

The hourly fee I've agreed to covers about half the time required to complete the scope of work, but at least it's enough to pay the month's bills. While an architect's fees may range between six and ten percent of the total building construction costs, a landscape architect's is typically a fraction of that, at about six percent of the landscape construction cost. So if the architect is getting about $20,000, the landscape architect earns about $2,000. Projects typically run over budget during the building phase, diminishing the budget for landscape even more. Clients ask for a "shrubbed up" parking lot, good enough to satisfy city code, or the "best bang for their buck" on a residential landscape. Something neat and tidy, maintenance-free, and cheap.

I shake the developer's hand and head north toward home.

The alfalfa fields of Algodones, north of Albuquerque, fill with acequia water, pooling in the low spots, spilling over at the ends of the planted rows. A tractor hums on the frontage road near the railroad tracks. The 80-miles-per-hour interstate borders the green growth and old adobe neighborhoods.

I pull into Santo Domingo Pueblo to gas up and see the steep red rock incline of La Bajada ahead, the extreme topographic transition from the 5,300-foot elevation of Albuquerque to the 6,960-foot elevation of Santa Fe. I make it back to Santa Fe in time to pick Olivia up from school.

We drive home chattering, sharing details about our day: my meeting in Albuquerque; hers performing a Spanish skit, followed by cooking and serving a hot lunch to classmates. Sometimes on our way to and from school we listen to music, usually of Olivia's choosing. This afternoon is one of those days; we're both in a lively mood, chipper about getting home to chillax. Speaker turned up, music Bluetoothed in, I hear the raucous guitar and gritty vocals of AC/DC's "Givin' the Dog a Bone."

It plays through once and I wonder what of the lyrics—"she's using her head, oh, she's using her head again . . ."—Olivia actually hears. I wonder if she's simply grooving to the beat and letting go, listening to rad rock with mom, or if she's taking in the words and meaning of "she's blowing me crazy," if she understands that the woman lowered herself to her knees, that she's given herself up.

"That one is so repetitive," I say, grabbing the device and skipping to the next song. "Let's listen to something else."

My hands tighten their grip on the steering wheel.

The song continues playing in my mind, in counterpoint to the other words echoing through my head. "I hate to see a woman beg." "Fruit is so messy." "We'll pay you more later." "Cut the old cottonwood down." "Biggest bang for your buck."

Once at home with homework and a snack in front of Olivia, I load up my new Mauser .308 and fresh ammunition into the pick-up. I strip off my skirt and blouse, pull jeans and a long-sleeved cotton shirt on, lace up my hunting boots. I need to sight my rifle in for the upcoming elk hunt. I want to go shoot, to see clearly, focus and take aim at a practice target.

I drive the fifteen minutes from our home across the highway to the nearest public land at the Caja del Rio, a series of piñon-juniper-dotted mesas and sandy canyons that lead to the Rio Grande. Piles of empty shotgun shells and ammunition cartridges, televisions obliterated by close-range target practice, broken glass, crunched Bud Light cans, and mini tequila, gin, and whiskey bottles reflect low fall sun. Old plywood, large plastic bottles, and discarded car hoods remain as makeshift bull's-eye targets. This national forest landscape lies broken by consumption, but it's the nearest place to take aim.

As I unload the shooting gear, I step over perfect piles of oval elk turds, some scattered into individual droplets, evidence of animals pooping in motion. Elk and deer cross here, following the draws from neighboring gated golf course communities where fertilized grass and manicured ponds provide sustenance. The elk hooves form pointy semicircles in scathed ground, press faded cans into stony soil.

I set up my target and shooting bench, put my cap and gloves on. With the rifle pressed firmly at ninety degrees from my shoulder, safety clicked off, I look through the scope. I see the red marker bull's eye on butcher paper stapled to the plywood I brought from home. I breathe in, exhale, pull the trigger, and say to myself, "This one's for you."

"OK, Christie," the contractor says on a voice-mail message, "You can have $3.68 a square foot. I think that's what we got away with on the other project."

After shooting six groups—three shots each at six different mini-targets—to make sure the rifle and scope are in line for a 150-yard shot, I pack up, fold the shooting bench into the pickup bed, and head home to Olivia.

On June 22, 2019, thirty-four years after Mamma, my grandmother, died, Olivia and I had harvested two bushels of sour cherries, one from each tree just north of our house. We froze them in gallon Ziploc bags to savor during the months ahead. Now, after the Mauser has been sighted in and Olivia's homework is complete, their maraschino-colored juice splatters my white shirt. One cherry plucked into the bowl, one popped into my mouth. I swallow, a tart pucker curling my tongue as the rubbery fruit glides down my throat.

We celebrated throughout the growing season with iced cherry-lime sodas on the porch, glossy sugared cherry glaze spooned over vanilla bean ice cream, and cherry juice bubbling from flaky empanada crusts, pinched closed by fork tines. Today, we're eating them straight. We pit the thawing cherries at the dining table, pushing the white crocheted cloth to the side, our flip-flopped feet sticky with juice squirting from our fingertips. We lap up the sour-sweet messiness.

"Mom," Olivia says, "Clayton and Bria have a little stand up the road between their houses. They've been selling lemonade there. Think they'd let me sell cherry-limes?"

"Bet so, sweetie."

"Wonder what they'd sell for? How much?"

"I don't know, hon. What do you think they're worth?" ○

Rob Jackson

WORDS

Knowing what to say is like the rain
that comes when it wants, then vanishes for days,
or like a ship that sails into port at last;
people celebrate its return with toasts and speeches,
knowing it will leave again.

Sometimes my words come easily.
I'm as steady as an anchor
on the evening news whose perfect sentences
float off in well-formed balloons
once the camera light turns red.

At other times the light turns red
and I can only screech
to a halt in my car, drowning in silence.
A few feet away, a homeless person
grips a scrap of cardboard,
trying to catch my eye: "Please help."
Silent, staring awkwardly ahead,
I look for change,
a sign of ships or rain
on the horizon.

Carlesto624   *The Blue Truck*
There are so many things in this life we take for granted. . . This truck is no exception to that rule. I wondered what kind of life this truck had and was it happy where it had arrived.

Natasha Balwit-Cheung

MAKING BUILDINGS
OUT OF PLACES

When my youngest sister was a few years old and I was a teenager, I'd tuck her into bed with five questions. "What is door?" she'd ask. "Wood, usually, and metal for the hinges—keep your fingers away from the crack," I answered, looking at the bedroom door. "What is fire?" she'd ask. I struggled. "What is wall?" she asked once, and I remembered something my father had told me about gypsum: it came from the shells of ancient creatures, mined from an inland sea long evaporated, east of where we lived.

Gypsum is a soft mineral composed of calcium sulfate and water. Processed into panes, it becomes prefabricated drywall. Heated and dehydrated, it is turned into a product known as plaster of Paris. Gypsum is used as an additive in the manufacturing of Portland cement, a primary ingredient of concrete. It also has agricultural uses, and has been used as a soil amendment for at least 250 years. Surface-mining operations in Oklahoma, Iowa, Nevada, Texas, and California cumulatively produce about two-thirds of the gypsum mined each year in the United States. The mines my father was probably thinking of when he gave me his poetic, geological definition of gypsum are the ones at White Mesa in Sandoval County, New Mexico. I am sitting in a room about forty miles from White Mesa, surrounded by rectangles of drywall covered in plaster, but it's not likely that this gypsum came from there.

Transport, production, and disposal (or "afterlife") chains of building materials have transformed wildly over the last century. A few hundred years ago, I'd almost certainly be sitting inside an assembly of materials sourced nearby. Now, there's no guarantee. Construction-material supply chains now spiderweb across the globe. An article in the *New York Times* on the effects of COVID-19 on construction says that for a project in San Francisco, California, "theater seats from Colombia, wooden doors from Italy and metal ceilings from Germany bound for a new corporate campus in Silicon Valley [were] likely to be delayed."[1] The developer of a 931-unit apartment tower in the Hudson

← *Lodge in Norris Park*, between 1933 and 1945
This is one of the outdoor recreational resorts developed by the Tennessee Valley Authority in cooperation with the National Park Service and the Civilian Conservation Corps (CCC) of Norris Lake, Tennessee. The lodge contains facilities for dining and dancing and has a beautiful view of the lake from the dining terrace. The park is developed in a rustic style native to the surrounding territory principally because the CCC program, which needed such opportunities, could furnish only low-skilled labor and practically no materials except those which could be procured by enrollees from local resources, such as timber and stone.

Yards area of Manhattan, New York, was set to install cabinets from Italy, but decided instead to purchase them from a New Jersey manufacturer, given Italy's national lockdown. The New Jersey manufacturer, however, depends on raw materials from China—so delays could not be avoided.

Historically, buildings in many ways reflected and resembled their respective surrounding landscapes. The limestone and marble building blocks of the Parthenon were not shipped across international borders, or even across their own country. The steel bones of the buildings of Hudson Yards might have been.[2] The megadevelopment—called a "billionaire's fantasy land" and "corporate city-state"— spans millions of square feet and represents the city's vast economic inequity. Its materials are products of globalization, energy- and emissions-intensive manufacturing processes, and transport over long distances. From the point of extraction or harvest of raw materials to processing, manufacturing, and distribution sites, building materials get pushed around the planet, eventually arriving at the construction site—and someday in the mournfully not-too-distant future, most likely, a disposal site.

My favorite introduction to embodied energy in buildings—"what is required to extract, manufacture, transport, and assemble a building's constituent materials"—comes from an entry by the architect David Benjamin in *Metropolis* magazine's Sustainability Glossary series. Benjamin writes,

The concept of embodied energy allows us to examine the intersection of architecture and the environment with a fresh perspective. By considering the production of a building as an act of energy expenditure, architects might start designing systems that extend beyond the boundaries of their walls and facades.

They might see buildings as temporary formulations of matter, energy, and labor that are connected to other formulations preceding and following the life of a building. And embodied energy might offer a framework for designing—or at least accounting for—many invisible yet essential aspects of the built environment, including embodied carbon, embodied water, and embodied labor.[3]

Consider an accounting of the attributes of building materials that includes labor, transport, energy expenditure, and land. Consider the fact that buildings and neighborhoods represent material resources, and choices about which materials to use and where to get them are those of consumption, which will be monumentalized for decades in a structure people use. Consider the radical possibilities of a future in which buildings again are made from materials that are regeneratively farmed, locally and responsibly sourced, and respected and valued both before and after their incarnation as components of a given building.

In urban environments, cities appear and are perceived as products of industry, with real estate as a surrogate for land. But even cement— one marker of human domination of land—is made from calcined lime and clay, dug from the earth before a tightly controlled (and emissions-intensive) chemical process transforms them. The immediate surroundings of human habitations are embedded in broader surroundings: the fields and quarries and forests that are the provenance of building materials are pieces of our larger homes.

In William Morris's essay "Art and Socialism," decency of surroundings is a necessity due rightly to all people, following the necessity of "honourable and fitting work" and preceding the necessity of leisure. Decency of surroundings, he writes, includes good lodging, ample

Grand Canyon National Park: Lookout Studio – Early Morning
Lookout Studio is one of the few remaining examples of the work of master architect and interior designer Mary Elizabeth Jane Colter. Colter's place in American architecture is important because of the concern for archaeology and sense of history conveyed by her buildings, and the feelings she created in those spaces. Her creative freeform buildings, Hermit's Rest and Lookout Studio, took inspiration from the landscape and the Ancestral Puebloan people, and served as part of the basis of the developing architectural aesthetic for appropriate development in areas that became national parks. In keeping with her style, Lookout Studio was carefully designed with native stone and an irregular roofline to blend into the rim of Grand Canyon.

space, general order, and beauty:

That is (a) our houses must be well built, clean and healthy; (b) there must be abundant garden space in our towns, and our towns must not eat up the fields and natural features of the country; nay I demand even that there be left waste places and wilds in it, or romance and poetry—that is Art—will die out amongst us. (c) Order and beauty means, that not only our houses must be stoutly and properly built, but also that they be ornamented duly: that the fields be not only left for cultivation, but also that they be not spoilt by it any more than a garden is spoilt: no one for instance to be allowed to cut down, for mere profit, trees whose loss would spoil a landscape: neither on any pretext should people be allowed to darken the daylight with smoke, to befoul rivers, or to degrade any spot of earth with squalid litter and brutal wasteful disorder.

Not only must our houses be "stoutly and properly built" to house people well and last a long time, he says, but also that "they be ornamented duly"—to bring joy and romance to the act of living in them. Overall, he posits, our houses and living should not degrade the land or the labor of the people.

There are modern designers working to comply with these demands. MASS Design Group, architects of the Rwanda Institute for Conservation Agriculture, excavated, harvested, or sourced 96 percent of construction materials for the institute from within Rwanda. Ninety-eight percent of the labor for the project was provided by people living within one hundred miles of the project site. The earthen material for compressed, stabilized earth

blocks and rammed earth walling was dug nearby after multiple sites were tested to identify the one most appropriate. The ceramic tile roofing was fired in kilns fed by coffee husks, a byproduct of local agriculture.

Other works are underway that serve as inspiration for careful architecture and construction informed by and respectful of land and its resources. Hive Earth, based in Accra, Ghana, makes rammed earth walls that are strikingly beautiful, "ornamented duly" with undulating waves and textures of layered earth mixtures. Co-founder Joelle Eyeson told *ArchDaily*, "In Ghana, we have so many different variations of earth that we can get light beige, gray, red or even black. Sometimes we add iron oxide pigments to create brighter colors if the

Carol M. Highsmith *Remains of the homes of ancient cliff dwellers in Canyon de Chelly (pronounced duh-SHAY), a vast park on Navajo tribal lands in northeastern Arizona, near the town of Chinle, that is now a US national monument*
Ancestral Puebloan peoples constructed the "White House" using sandstone, mud mortar, light-colored plaster, and wood.

client asks for it."

Hive Earth can build a one-bedroom house this way for five thousand US dollars, and it will keep cooler than an equivalent house made with more cement or other materials. They are sound- and termite-proof, and free of harmful chemicals commonly present in cement.

Another place to intervene in a building-material supply chain is at its end. In 2016, Portland, Oregon, banned demolition for houses more than one hundred years old, requiring "deconstruction" instead. Alisa Kane, Portland's climate action manager (then the green build-ing manager in the city's Bureau of Planning and Sustainability) told me they narrowed in on the buildings the deconstruction policy would cover by looking at which homes were being demolished, where they were in the city, and what they were made of.[4] They found that older homes and the materials in them were concentrated in certain areas. As Portland grew, it expanded outward, in rings, with the houses that ring the central city tending to be much older than those farther out. Those older homes are the ones, she said, with "great old fixtures, clawfoot tubs, old beautiful doors, and then the lumber inside their walls—that's our old-growth forest."

"As you drive out to the coast, you see the clear cuts that have happened over time. Those are our forests in our homes," she said. The deconstruction ordinance has bolstered the market for that wood for salvage and reuse, and has supported the development of a decon-struction workforce that serves as a pathway into the trades.

In this way, the building economy of Portland becomes more circular, and the mate-rial richness of the existing building stock gains increasing visibility. Architectural modernism obscured the relationship between buildings,

materials, and landscapes, but policies and designs can poke holes in that obscurity. The hope invested in new biomaterials—bricks grown from fungi, living building materials engineered from lichen and cyanobacteria, and others—and old ones, including straw bale and sheep's wool insulation, reveals a hunger for buildings that exist as extensions of, and complements to, the living systems of their environments. It's a reasonable desire, and we deserve to meet it for ourselves and our world. With a holistic understanding of material flows and supply chains, designers and builders (and planners and policymakers and people engaged in grassroots efforts to shape the built environment) can reach toward this aim. The slow food and slow fashion movements prize local economies, attention to raw materials, and systems contrary to the churning exploita-tion of dominant modes of production and consumption. Slow architecture can prize those things, too. ○

Notes

1. CJ Hughes, "Chinese Copper, Italian Marble: Coronavirus Shipping Delays Hurt Developers," *The New York Times*, March 20, 2020.
2. The May 2020 "Global Steel Trade Monitor," published by the International Trade Administration, notes that the US imported 26.3 million metric tons of steel from about eighty countries and territories in 2019.
3. David Benjamin, "Sustainability Glos-sary: Embodied Energy", *Metropolis*, January 14, 2019.
4. Alisa Kane, telephone conversation with the author, October 2016.

Rose Robinson

WALKING THE CAT

He does not know where the catnip comes from
and he has never seen the rain without a window.
We can see the bones of his hips through his fur,
so after sixteen years that poor old cat
can go outside for U-Pick Ganja and
when it begins to drizzle I stay outside with him.
He moves so slow that I hold him for a while and
as we walk I teach him the names of the trees
and the flowers I know. Baby, that's a birch,
and buttercups. These are pines, and violets,
and love, that's a locust. Because to me
he's like some city kid who doesn't know
how tomatoes grow. And I've spent my life
picking yellow coyotes and beefsteaks
and I knew where our thyme and chives grow
before I knew what the herbs were called.
If only I could hug his hands into soil,
wrap his arms around trees, and teach him
to tend to the earth and sing to plants and
make up for all those years he spent inside.
But all I can do is sit him in the crook
of an apple tree he is too old to climb.

Amy Lindemuth

BUILDING RESILIENT URBAN FORESTS

A Conversation with Green Seattle Partnership's Michael Yadrick

In the mid-1990s, aggressive invasive species and limited maintenance resources severely impacted Seattle's forested parkland. Public agencies and private organizations recognized the urgent need to address the health of city forests and the concerted public-private partnership needed to tackle this colossal effort, and in 2004, Green Seattle Partnership was formed. Managed by Seattle Parks and Recreation, this partnership is the largest urban forest restoration project in the country, with the primary goal to restore and maintain 2,500 acres of forested parklands by 2025. Since its inception, land acquisitions have brought the total acreage to more than 2,750 acres. The partnership's twenty-year strategic plan calls for the establishment of financial resources to provide maintenance and secure the ability to sustain the forests in the long term. Cultivating an enthusiastic and involved community of volunteers and leadership is another basic outcome identified in the plan.

Now in its fifteenth year, the partnership has embarked on several research and data collection efforts to assess restoration efforts and inform proactive efforts to address climate change, increase urban density, and ensure equitable access to healthy, thriving open spaces. In June 2020, I spoke with Michael

Yadrick, one of the partnership's plant ecologists. Michael is leading several on-going assessments and collaborations with regional partners testing alternative strategies to restoration efforts.

How have the realities of climate change and continual population growth in the Puget Sound region influenced the partnership's approach over the years?
Over time we're doing more long-term stewardship and maintenance. We were, originally, about building tree canopy, but natural areas are already 90 percent canopy. We're removing threats from the restoration and re-greening with a diverse set of native trees and other plants.

Climate change is huge. Even nine years ago, it seemed way off in the future. Now more and more we know what the data says and we're seeing impacts and species decline in certain areas. We're exploring efficient ways to adapt quickly to changes in the climate, like the tree sourcing [from areas outside the Puget Sound]. The key to this approach is whether these species can "hang" in our area. When planted trees eventually sexually mature and mix with the local population, we can see if they can genetically mix and survive. We have models to look to, but not many people are doing this work in urban areas. Besides the technical fixes, there's [a need for] building the social capacity in the Parks and Recreation Department. Building contingency funding for unknowns, like landslides. Landslides are interesting because if they're left as is, invasives re-populate the slide area. If we're proactive, we can re-forest slide areas and reset the restoration trajectory with the existing forest.

How does the partnership cultivate a volunteer base that is inclusive and representative of the communities where you work?
Volunteers have provided a wealth of local expertise and stewardship capacity. Some people tend the natural areas, like gardens! We have goals to promote participation that mirrors the demographics of Seattle's neighborhoods. While we are traditionally white-led, we acknowledge racial and social justice are key to keeping the restoration effort relevant. We do this by inviting people to take the lead on restoration in their neighborhood park, while also partnering with organizations who have access into communities that surround our forests.

Have any outreach processes or tools been more successful than others to increase inclusivity and diversity in the partnership's projects?
We definitely rely on face to face interactions a lot. What we've generally relied on, our forest stewards model, is split between retirees and young professionals. The retirees have been with us for a long time and are our strongest advocates. There's probably more ethnic diversity in the younger group. We intentionally seek out nonprofits who have access to the community and reflect the community. For instance,

Tilth Alliance works with Ethiopian elders in southeast Seattle neighborhoods. That relationship rose out of Tilth's location in the Rainier Valley. We supplement funding to give stipends to the elders for their work. The Student Conservation Association has also helped recruit and support diversity.

How does the partnership's mission support social justice and equity in local neighborhoods and open spaces?
Racism has left a signature on the city's neighborhoods and unfortunately, there are parts of town that are lacking forest canopy and its associated benefits. The outcome is that certain places are hotter and more polluted than others. The remarkable thing about the program is that we work citywide. Not only are we stewards of the Olmsted parks [originally designed and/or conceived by John Charles Olmsted, Frederick Law Olmsted's son], but we also currently work in or have plans to serve every natural area that has not been cared for in over one hundred years. We also recognize past hurts and harms by engaging directly with communities such as the Duwamish Tribe and the United Indians of All Tribes Foundation in the West Duwamish Greenbelt, as well as Daybreak Star Indian Cultural Center.

What approaches have successfully engaged future leaders and forest stewards?
One of the coolest things is that the program is embedded within Parks and Recreation. We join forces within the department and with community partners to integrate youth leadership development, green job mentoring, and environmental education in our programming to elevate diversity within the restoration effort.

→ **Seattle Parks and Recreation** *Youth Program*

What has been one of the most successful community partnerships you've been involved with? What were the ingredients that made it successful?

Almost every nonprofit partner has a partial focus on engaging schools and youth. Youth Green Corps has probably been our longest running program. The program sits within the Seattle Parks and Recreation Trails Program and the youth involved carry out restoration alongside the major trail maintenance projects. Several of the youth have been picked up as seasonals, or part-time positions, at Seattle Parks and Recreation—the end goal would be employment with us or somewhere else in the green-collar industry.

What connections do you see between the work the partnership is doing to restore urban forests and the health of regional wildlands and farmlands?

Our region of Washington in the Puget Trough is home to four million people. The forests and farms beyond our cities are heavily influenced by what happens in the metropolis and vice versa. What makes this place livable is our access to a sustainable local food system and places where wild things can thrive. Also, our land ethic and practices, particularly in the city, has an incredible impact on the health of our creeks, rivers, and Puget Sound.

In March 2020, you wrote a Green Seattle Partnership blog post about Climate Ready Urban Forests. What are Climate Ready Urban Forests and what actions is the partnership taking to realize them in Seattle?

Climate readiness means we preserve as many forested natural areas as possible so the land becomes something else. Then, we assist the recovery of the urban ecosystem through changing environmental conditions. It is a process by which we educate ourselves and train for the new normal so we can move beyond the discussion about the impacts of climate change and focus on equitable

Ray Filloon, USDA   *Keen Ponderosa Pine Tree Classification*, Fremont National Forest, Oregon

adaptation. As climate change accelerates, everyone is going to be under pressure to implement projects on relatively quick time-lines that are inconsistent with addressing underlying issues of structural violence. Socially just adaptation requires inclusion of peoples who are vulnerable to climate impacts, particularly women, the elderly, youth, BIPOC, and LGBTQ individuals.

To date, the partnership has chosen assisted gene flow in lieu of assisted migration[1] to help combat the tree species morbidity associated with climate change. What makes assisted gene flow preferable to assisted migration in the projects where the partnership is implementing this approach?

Seattle is situated in the middle of the habitat range of several native tree species. That provides us with ample options to move genotypes of species to the Puget Sound from provenances located south of us. Personally, I am interested in other species from outside the range that are drought tolerant and potential "winners" through changing climates. Some grow quite well in our wider landscape.

I've wondered what our obligation is to contin-ue supporting other species in the ecosystem that typically depend on our native trees. Would you mind sharing your thoughts on this topic?

Where native species cannot replace ecosystem functions, we could choose to actively assist colonization of species outside of their current range. Some candidates are ponderosa pine, incense cedar, giant sequoia, coast redwood, or even golden chinquapin. There is a semi-local population of ponderosa pine down near Joint Base Lewis-McChord in Pierce County. With the decline of Pacific madrone, some botanists think chinquapin may be a good broadleaf

evergreen surrogate. There is some concern that sequoia may not develop viable seed outside of its native range, though it establish-es well in the Puget Trough.

There are unresolved trade-offs in pursuing "light-touch" versus bolder climate-targeted "renovation" type activities for sure. Some of this comes down to some existential questions about historical fidelity and our willingness to accept risk for future generations. There is a lot of research, though not many people are dialed into the science and there are lots of "feelings" to navigate. It's a conversation that profession-als need to be having, and we need to bring the public along with us without leaving people behind in the discussion.

As the last five years of the twenty-year strategic plan sunsets, what is your hope for the future?

While some of the intense restoration work will diminish over time, there will always be a need for long-term forest management. We live with a legacy of land use activities that started in the 1800s and was not conducive to growing forests. It took decades for non-native invasive species to establish and disperse across our natural areas. Twenty years is a quick turnaround. We are still ambitious, and I expect we will have to pivot to keep individuals engaged in new ways and with new priorities as the city changes. ○

Note

1. Assisted gene flow is the translocation of pre-adapted individuals to facilitate adaptation of planted forests to climate change, and assisted migration is the intentional translocation or movement of species outside of their historic ranges in order to mitigate actual or anticipated biodiversity losses caused by anthropogenic climate change.

Nelly Nguyen

CHUỖI TÁC PHÂM THỤC-NGHIỆM
FOOD LAB SERIES

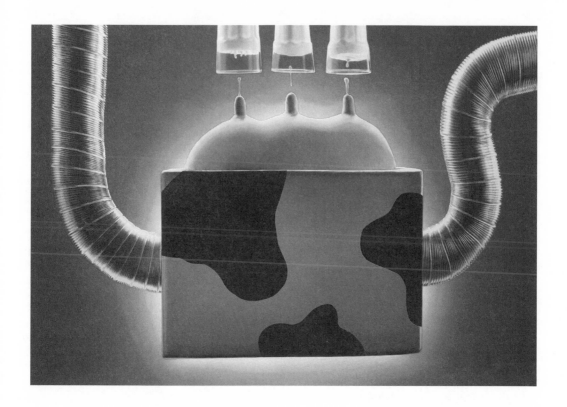

Photos by Briana Olson

Rob McCall

FULL CORN MOON IN MAINE

Natural Events—
Looking out in the early light of a damp morning,
we are likely to see spiderwebs bejeweled with
dew, strung with droplets refracting white light
into rainbow colors like cut crystal. These webs
may be spread over the grass or hung from the
eaves or branches of trees. They are made of a
fine fiber produced from the spiders' own bodies.
The little webs spread across the grass like tiny
trampolines are made by the Agelenidae, also
called grass spiders or funnel spiders. They hide in
the funnel and wait for their prey to get caught
in the web, then spring out and wrap it in more silk.
The large webs on the porch or under the eaves are
built by members of the Eratigena family. Females
of all species are loading up on flying insects to
give them the protein they need to lay lots of
healthy eggs that will survive a tough Maine winter
to hatch next year. One year we watched their
Kabuki-like mating ritual which was beautiful yet
horrifying as the male was devoured at the end.
Maybe he thought it was worth it. And, as any
reader of *Charlotte's Web* knows, she will soon die
too. For the arachnophobes among us, it's good to
remember that these spiders eat masses of harmful
insects and are a net benefit to the environment.

This was originally published in Awanadjo Almanack *for the week of*
August 28–September 4, 2020. The Awanadjo Almanack appears in the
Quoddy Tides, The Weekly Packet, *and* Maine Boats, Homes & Harbors,
and as a podcast on WERU *Community Radio.*

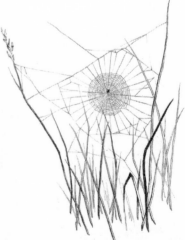

Melanie Gisler, Maria Mullins, and William Hutchinson

DEVELOPING NATIVE PLANT MATERIALS FOR ROADSIDE DUST MITIGATION IN SOUTHERN NEW MEXICO

Airborne dust has resulted in extremely hazardous road conditions along many highways in New Mexico, including an area in the southwestern part of the state where Interstate 10 bisects the Lordsburg Playa. Playas are shallow, flat, dry lake depressions that form when water drains into a basin with no outlet and then evaporates, leaving a deposition of salt, sand, and fine soil. The Southwest Office of the Institute for Applied Ecology (IAE) is currently working with the New Mexico Department of Transportation (NM DOT) and the New Mexico Bureau of Land Management (BLM) on a project to mitigate dust through improved plant cover while conducting a study to improve our understanding of native species and identify plants that are able to survive saline soils, drought, and other harsh conditions of the playa.

In early 2018, IAE researchers began by assessing Chihuahuan Desert plant communities to identify plants that might be able to withstand these conditions and also mitigate erosion and dust. Then, through extensive literature reviews and comparative species-rankings, IAE and an NM DOT Technical Advisory team identified grass, forb, and shrub species that are drought tolerant and quick-establishing, and show promise for restoring and controlling

dust-generating soils in playas and surrounding habitats in southern New Mexico.

One of the goals of the project is to establish species not previously used in roadside restoration projects in these challenging habitats, and to increase the regional availability of such species. An equally important goal is to cultivate local sources of frequently used restoration species, including perennial grasses such as side oats grama (*Bouteloua curtipendula*) and alkali sacaton (*Sporobolus airoides*). Studies have repeatedly demonstrated that plants adapt to the local environment and that locally sourced seeds have higher levels of both short-term and long-term success.[1] The study also provides the opportunity to compare locally sourced seed with the same species from commercial, nonlocal sources. In addition, our seeding experiment will present a side-by-side comparison of these common restoration species with novel species that are not yet commercially available such as golden crownbeard (*Verbesina encelioides*), Indian rushpea (*Hoffmanseggia glauca*), and slender goldenweed (*Xanthisma gracile*).

During the 2018 and 2019 field seasons, seed collecting field crews wild collected seed from local sources and purchased commercially available seed from large seed producers. In

July 2020, seed was sown in research plots at five sites near the Lordsburg playa. The study will test which seed sources, species, and seed mixes are most effective at establishing in these challenging soils and which are most effective at preventing car accidents and other public safety hazards that result from dust storms. What we learn will help the NM DOT enhance revegetation efforts while actively restoring soils in dust problem areas around the Lordsburg Playa. It may save lives. Determining which germplasm is most successful will improve revegetation practices, encourage growers to produce these native plant materials, and share best seed practices among restoration practitioners and other seed users.

This research aligns with the mission of the Southwest Seed Partnership (SWSP), a collaborative effort to increase the availability of appropriate native plant materials in New Mexico and Arizona. IAE coordinates the SWSP and supports native plant research by providing infrastructure through their regional seed collection program. One goal of the partnership is to build a native seed industry in the Southwest, and research such as this dust mitigation project will provide valuable information supporting this effort. ○

This five-year, multi-objective project is funded by the Federal Highway Administration, the Bureau of Land Management, and the Native Plant Society of New Mexico. Landowner partners include NM DOT, New Mexico State Land Office, and local private landowners such as Rafter JL Ranch and Kerr Ranch.

Note

1. J. Joshi, B. Schmid, M.C. Caldeira, P.G. Dimitrakopoulos, J. Good, R. Harris, C.P.H. Mulder, "Local adaptation enhances performance of common plant species," *Ecology Letters* 4, no. 6 (2008): 536–544. See also M.J. Germino, A.M. Moser, A.R. Sands, "Adaptive variation, including local adaptation, requires decades to become evident in common gardens," *Ecological Applications* 29, no. 2 (2019).

Dorothea Lange *Dust storm near Mills, New Mexico, 1935*

Michael McMillan

A COLLECTIVE DESIGN

Several years ago, during my dreadlocked dirtbagging days, I hitchhiked down to Mount Princeton Hot Springs in the San Isabel National Forest with the intention of solo hiking sixty miles of the Colorado Trail to Monarch Pass. While soaking, I met a worn-out, bearded backpacker probably twice my age. He had just gotten off the trail, was also solo hiking, happened to be a mycologist, permaculture enthusiast, and botanist—and was also named Michael, spelled Mycol.

Travelling together happened serendipitously, our conversation and our feet wandering off into the mountains. Equipped with my Plants of the Rockies field guide, I stopped to eagerly identify every wildflower, cactus, shrub, and tree I could find. Mycol seemed thrilled to have a youngster to mentor along the journey. Instead of painstakingly identifying each and every plant from the book, he taught me patterns to look for in flower and leaf structures. Each day we'd play the trail-rousing game of "name that plant family." Before long, I could name almost every plant family in the forest. Asteraceae (sunflower family), one of the largest families on earth. "If the flower looks like a star, it's probably an aster," Mycol said. Brassicaceae (mustard family), also known as cruciferae for their characteristic "cross" flower structure, always have four petals at ninety-degree angles. Fabaceae (pea family), known for their unique, sailboat-like flower structures and their relationships with nitrogen-fixing bacteria who reside on their roots. On and on we went, Mycol and I, botanizing all the way.

I wanted to be able to survive in the woods simply by relying on my knowledge and my relationship with the land. Learning individual plants and plant families was essential, but while walking, I began to notice the patterns of plant communities, or guilds. The brassicas, like watercress, tended to grow close to creek beds, under praying western monkshood, willows, and Doug-firs. The lupines, purple asters, and aspens emerged into previously disturbed areas, perhaps where there had been a rock-slide or an avalanche. Heracleum cow parsnips filled the understory of older aspen forests. Mixed with sweet cicely, red elderberry, columbine, and Fendler's meadowrue, the slow growing blue spruce worked its way towards the canopy as the aspens became old and died.

This is a practice of reading the landscape, and tuning into the ecological story that is unfolding before you. All of this must be considered when designing a space.

"Landscape design," I say, even though I feel like this phrase doesn't come close to explaining what I do. Depending on who's asking, I might dare say "ecological design." I have yet to come up with a title that encapsulates what I do. I strive to be a living systems enabler, to facilitate relationships between people and the land to bring about healing and abundance for all of life's biodiversity, down to the last soil microorganism, while maintaining humility and acknowledging that there is a lifetime of learning ahead.

Four years ago, I moved to a new community in southwest Colorado to work as an Americorps volunteer supporting a school-to-farm program. I fell in love with the area and cultivated a passion for growing food. I found mentors who could teach me the nitty-gritty details of growing all of my favorite vegetables in the harsh high desert climate. After a year and a half, I was hired to teach the agricultural science program at the middle school. My students and I grew blue corn, Navajo pumpkins, red curry squash, lemon cucumbers, and Walla Walla onions bigger than the sixth graders' heads. We cared for a heritage orchard of sixty-seven trees of nine varieties, pruning every winter with the local apple gurus. I'd start

each class with a Kichwa greeting, "*Ali mashi kuna*," and some students would reply, "What's up?," "Hey, Mr. Michael," and some "*Yá'át'ééh*," a greeting in Navajo. It brought me endless joy to send food-insecure kiddos home with bags of fresh vegetables for free. Before going to harvest, we'd read Robin Wall Kimmerer's guidelines for the "Honorable Harvest" aloud as a class: "Ask permission of the ones whose lives you seek. Abide by the answer." Some of the same students returned to my class every fall. After three years, we began to witness the rhythms of the land. We recognized the same family of finches returning each spring. We watched as cattails and cottonwoods moved into the pond we'd dug three years before. The land was teaching us, one season at a time.

"Will you help me get my garden started? I can pay!" a friend said one late summer day. Excitedly, I went into her backyard to see sparse grass, rocks strewn everywhere, overgrown box elder trees, and lots of potential. I dove in, first using the scattered river rocks to build raised beds. I planted clematis vines, raspberries, daylilies, and grass from seed. I put wildflowers in terraced garden beds coming down the slope from her back door. I re-leveled the flagstone patio, pruned the trees, and restarted her defunct compost bin. She worked with me some of the time. We talked about lots of things, from work and love to politics, music, and travel. Pulling weeds is one of the

Asteraceae Flower Family

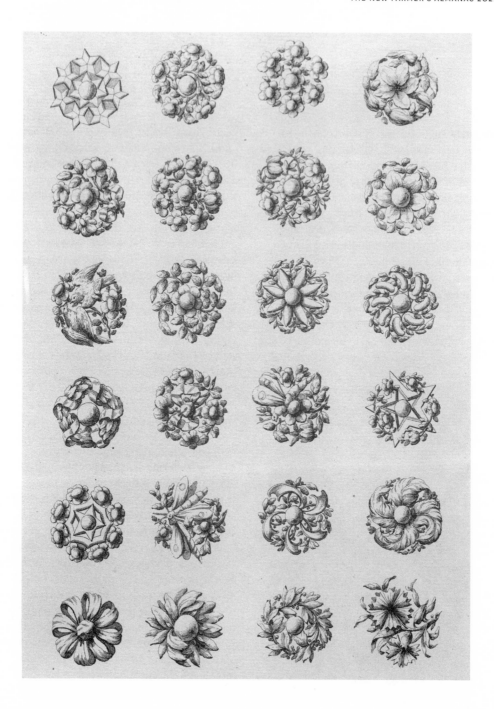

best ways to get to know someone; iced tea, lemonade, or a cold beer speeds along the process nicely.

Another friend asked me to help her start a garden, then another. Maybe I should start a business, I thought. Now it's been a year since I founded Solstice Sown Designs LLC. I've gotten lots of different gigs: designing a drought-resilient forest garden for a friend's farm, supporting a three-hundred-acre cattle ranch with custom irrigation, planting trees here and there, revitalizing a permaculture property that had been taken over by perennial pepper weed, and renovating a historic apple orchard with my trusty Stihl chainsaw.

All of these smaller, temporary projects brought meaning, new challenges, and new relationships to my life. Then my dream project sprouted up in fall 2019: designing a layered forest garden with edible, medicinal, nectary, and nitrogen-fixing species from ground cover to canopy.

Visualizing the garden-to-be, I saw strawberries running over lichen-covered logs and rocks, violets painting touches of purple across the ground. As the seasons change, each organism plays its part in the ecological orchestra. Daffodils, hyacinths, lilacs, tulips, and wild iris open the symphony. Spring deepens into summer and the bees take their time to visit each blooming flower: arnica, echinacea, lemon balm. King stropharia mushrooms silently decompose the wood-chipped pathways. Western monkshood delphinium stands tall and dark, growing around brambles of raspberries and blackberries. In the good years, I imagine there will be too many apples, pears, wild plums, serviceberry, and Nanking cherries for one family to eat. The leaves of tall aspen trees quake in the wind.

A family of Lewis woodpeckers watched me

as I went back and forth digging, planting, and watering to bring this vision to life. I can invite a multitude of diverse species to the biotic garden orchestra. Some will have an outstanding performance, some won't play at all, and others will show up unannounced. I've adopted a teaching from the Wild Yards Project which advises designers to envision when the thing you planned is long gone and the thing you have is much more than you hoped for and way past being yours at all.

Ecological landscape design isn't a one-and-done kind of deal. Developing rich soil full of humus and growing forest gardens takes years. It is a transformative and educational process aided and informed by science. I've learned that growing in my design practice is really growing to become a more powerful listener. I've learned that ecological design is a kind of dance with the landscape, continuously learning, adapting, and evolving as the land does.

The most successful work happens when my clients and I invest in a shared vision together. We both understand and accept that nature is the real "designer"; we simply observe and interact. As humans, we have the opportunity to become ecosystem enablers. Getting our hands in the soil and planting trees—and acknowledging that they will live much longer than we will—tends to deepen human relationships, too.

As I finish planting the mountain mahogany, I hear a call from the house, "Hey Michael, you've been working awfully hard, would you like some cookies and tea?" We end up sharing poems, land ethics, and stories—and that is all part of the holistic process that creates the relationship between land and people. ○

↑ *Attention boys and girls - $250.00 in prizes - miniature home building contest,* May 3, 1940
→ **Cassi Saari** *Flowering big bluestem (Andropogon gerardii) in a tallgrass prairie restoration*

OCTOBER

Remediations

National Parks, State Parks

Colleen Perria

THE ART OF OPENING OAKS

Restoring Oak Savanna in the Great Lakes Bioregion

Chin against bark, I look straight up the black cherry to gauge her subtle lean. The motor of my chainsaw breaks the quiet of the land. Sawdust mixes with the season's first accumulation of snow; it flies onto my coat and catches between work boots and wool socks. The light is fading and the whine of the saw cuts out. I remove my earphones, step back, wait, and listen as the black cherry tips towards the earth. Others working in the area idle their saws or turn from the brush fire to watch. Her winter buds brush against neighboring trees. The sliver of wood left at the base, attached as a hinge to control the fall, squeaks as it contorts, then breaks. The top of the tree bounces a little on impact, and the heavy trunk settles. Leaning on the new stump, I can see the gap of sky where the cherry, for thirty-some years, drank sunlight.

Now the upward curving branch of a black oak casts the only shadow on this particular patch of ground. The oak is around sixty years old, though it is the same size as the younger cherry beside it. This oak branch will soon stretch east, occupying the felled cherry's space.

Our shoestring crew works year-round along the banks of Iron Creek, in the River Raisin watershed of what is now called southeast Michigan. Equipped with finesse behind the chainsaw and lean muscles from hauling brush, we labor to restore the oak savanna ecosystem that was once common but is no longer. The keystone species of this system are oaks and humans.

Open-grown oaks are trees graced with ample space and sunlight. Oaks in a savanna need not strain or compete with fast-growing box elder, black cherry, or invasive shrubs. Instead, these oaks have room to spread their limbs, swoop, curve low, and convert the sun and earth into many plump acorns. They have a particularly healthy form and are inviting trees to climb. The oaks that are our neighbors here are black oak (*Quercus velutina*), white oak (*Q. alba*), red oak (*Q. rubra*), and burr oak (*Q. macrocarpa*). Crouching for balance, these trees can weather any storm and grow for over four hundred years.

In supporting wildlife, their value cannot be overstated. As Doug Tallamy writes in *Bringing Nature Home*, "Since the demise of the American chestnut, oaks have joined hickories, walnuts and the American beech in supplying the bulk of nut forage so necessary for maintaining populations of vertebrate wildlife. Acorns filled the bellies of deer, raccoons, turkey, black bear, squirrels, and even wood ducks."[1]

Additionally, Tallamy writes, oaks feed the

← Ernst Ferdinand Oehme *Study of a Tree [oak]*

less-visible creatures of the woods: "What we have under-appreciated in the past, however, is the diversity of insect herbivores that oaks add to the forest ecosystems. From this perspective, oaks are the quintessential wildlife plants: no other plant genus supports more species of Lepidoptera, thus providing more types of bird food, than the mighty oak."

To understand the open-grown oak is to understand the oak savanna ecosystem, also called oak barrens or oak openings. In the story of savannas, the main character is fire. For at least the past six thousand years, fire has painted this land often.[2] In spring and fall, parts of the landscape were ablaze. Now seldom seen, black was then common. Flowers, grasses, and small shrubs turned to ash, seeped in and enriched the soil, or blew away in the wind.

Following the burn came many shades of green. The curls of big bluestem grass pierced the surface from below and lobes of hepatica grew at the base of oaks. Each spring, regardless if the ground had been scorched or not, green crept in slowly, then covered the world. The fall brought the blend of tan, brown, gold, purple, red, and the whole palate of autumn. Spring or fall, as fire moved once more across the land in search of brittle leaves and dry grass to consume, the earth again turned an ashy black.

Humans set these fires. Lightning strikes—sparks from colliding rocks and other intersections of ideal circumstances—occurred to ignite the tinder, but it was the resident humans of the southern Great Lakes region that most often moved fires around. Why did the Potawatomi, Sauk, Fox, Kickapoo, Miami and others here regularly set fire to the land? The answers are many: hunting tactics,

renewal of plants, maintaining garden beds, the pleasure of navigating a burned and open landscape, and to intercept the lightning fire and tend a prescribed fire before a wild one swept through. I suspect another reason is the feeling of exhilaration and awe that ignites in us when we humans set a prairie ablaze. People prescribed fire often and for so long that the land developed characteristics and species particular to the savanna. Savannas host plant communities that are specific to their eco-systems, and so also host the moths, butterflies, and other insects that require these plants to feed and raise their young. Savannas are not only born of fire, but maintain their existence through periodic fires.

Oaks are the bones and structure of a savanna. They have extremely thick and fire-resistant bark. Fires that would light up the oily needles of red cedar (*Juniperus virginiana*) might feel harmless or even fulfilling to a mature oak. Oak leaves are crafted to carry fire very well, with large lobes that dry at various angles, allowing air to circulate. This characteristic is not ubiquitous among foliage; fallen oak leaves' ability to dry quickly and burn thoroughly is superior to all other neighboring species. Oak leaves also decompose slowly. They are ready to burn even after a winter's worth of Michigan snow has pinned them and drenched them. Oaks are ready for fire and they accommodate fire; perhaps they desire fire.

When savannas were first burned into existence thousands of years ago, trees that couldn't take the heat were thinned out or inhibited from growing in the first place, resulting in a landscape dominated by various grasses, flowers, alpha oaks, and beta shrubs. Neither woodland nor prairie, instead a blend that encompasses elements of each, oak savannas were diverse, enchanting, and unique. Butter-

flyweed, whorled milkweed, wild lupine, black-eyed Susans, asters, goldenrods, wood betony, bergamot, pale-spiked lobelia, big bluestem, little bluestem, and indiangrass soaked up the sun. Black raspberries in the low areas created fire barriers to let wild plum, hawthorn, saskatoon, hazelnuts, and New Jersey tea thrive. Towering over this plenitude, letting soft dappled sunlight in through their far-reaching branches, stood the oaks. This was the southern Great Lakes bioregion: southern Michigan, southern Wisconsin, Minnesota, Ohio, Illinois, Indiana.

But they are no longer. By the 1700s, violence done by white settlers, soldiers, and government policy—to Indigenous peoples, to the creatures on land and in water, to the land itself—was widespread. Ignorant of the ecology of this land, white settlers feared and suppressed Indigenous use of fire. They operated with fences, permanence, rows, lines, plots, borders, ownership, and the supreme sense of entitlement that has followed Western civilization wherever it spreads. As land grabs took place and fire-lighters were evicted or murdered, one of four things occurred:

1. The lush and open savannas were ploughed up and planted with wheat or other managed crops, and the oaks milled or split for firewood.
2. Seedlings of trees that previously could not coexist with fire sprouted prolifically, creating dense young forests.
3. Savannas were grazed heavily until conservative plants no longer sprouted and European grass and garden weeds dominated.
4. There stand our metropolises, paved, grey, smelly, contrived.

In *Reading the Forested Landscape*, Tom Wessels notes that the coastal savannas of New England disappeared within fifty years of white invasion and fire suppression.[3] Memory shifts: Although the ecosystem had been a savanna for ages, regional lore is of deep, primeval forests.

Today less than one percent of what was Michigan savanna still exists. In *A Field Guide to the Natural Communities of Michigan*, the authors note that for two of the six categories of savannas, oak openings and burr oak plains, no places remain to provide even an example of these ecosystems.[4] They have been extirpated from the state.

Some desire to heal this open, bleeding wound. Our crew in southeast Michigan is debating, reading, and working with lungs full of smoke, sore backs, numb toes, frozen fingers, and thighs riddled with thorns to restore oak savanna to land that, for at least three hundred years, has endured the plough, fire suppression, logging, building, and competition with aggressive non-native species. It is a humble attempt at reciprocity with the land that hosts us. We find clues that pieces of savanna remain. It is still possible to connect the broken fragments—the seeds are still viable, the parent oaks still leaf out.

We cut the trees that surround oaks and burn the leaf litter that falls in the shadow of each oak to restore the process, restore savanna, and restore ourselves to a relationship with the land that gives so much. This work is an attempt to balance the take, take, take interactions between industrial civilization's dominant culture and the land—a worldview revealed through the insulting term "natural resources." Oaks of twenty to two hundred years must be flooded with light and allowed to stretch as far as they please.

Culling the trees that surround an oak is not work that we take lightly. We avoid formulas and textbook prescriptions—"cull 35 percent of non-quercus species in plot x," "leave a thirty-foot perimeter of cleared area from crown to crown"—and other such impersonal intellectualization. When people with different points of view, priorities, information, and understandings debate how to be a steward to an area, those differences will be reflected back in the land. A tree is a creature that provides shelter, food, and beauty, and is connected through millions of microbes to all other roots in the earth. A tree exists, also, for their own undivulged purposes. We cut when we have consensus that, to the best of our understanding, the benefits outweigh the cost. No one person condemns a tree. We go slowly. We spend time in an area with no particular agenda. We get to know a place when we walk, play, climb trees, forage, observe, and consider. Then we talk, question, debate. We mark possibilities, mark decisions, and finally, cut and leave our trace.

We attempt to balance the time it takes to weigh the consequences with the urgency felt by the sun-starved bottom limbs of the oaks and our responsibility to the seeds of savanna species waiting in the seed bank of the soil. They are still viable, ready and waiting, but their biological clock is ticking. Fifty years may have passed since the wood betony bloomed, was pollinated, and buried her seed in the earth to wait for the proper conditions to sprout in. It may have been that long since the seed of whorled milkweed was carried by the wind and planted by the rain on the slope that faces

northeast. It has been waiting for sunlight and fire to signal that it is time to germinate. I have heard that one hundred fifty earthly species go extinct every day.[5] To not cut is also a choice with ramifications and results. Our friend fire also decides for us and may scorch a tree's base so that it dies during a prescribed burn, opening up the canopy and letting light onto the ground.

Here is a sketch of our crew's decision-making process: Should we alter this area? Do we know it well enough to change it? Which trees should come first? Cut this red oak for the sake of that white oak, you say? Is it cold enough? Has the sap stopped running? What is the report on oak wilt in the area? If we cut that hickory, how far away is the next? Will the seeds of the savanna flowers sprout up with the new light, or will the invasive seeds that are also dormant germinate first? Do we have enough wild seed to sow here? Will the bittersweet vine spread after the fire? Should we cut this cherry simply to collect and burn the bittersweet seeds tangled in her branches? The elms are struggling—should we leave this healthy one despite it shading this oak? Sassafras have such a dense canopy—should we be more aggressive in cutting them? How could you cut sassafras? I love sassafras! Can we connect this open area to that one there by cutting these trees? Will creatures respond better to a mosaic over a grid? Can we haul all that wood out if we cut it? Does this landscape flow? How can we break up the industrial lines? Would it be better to girdle this tree and leave the standing snag for the woodpeckers and chickadees? What can be thinned to soften the fence line between forest and prairie and blend it into a savanna? I know apples aren't native, but where are the hawthorns, crabapple, and dogwoods to fill the understory niche of

Chris Hoving *Smoke in the red pines*

savannas? Plus, there are great apples on that tree. Also huge morels grow there, I don't want to hurry the decay process ...

While considering these questions that guide us, we must remember that we are small, our awareness is still maturing, and we will never know all the complexities of the wild— nor should we. Management is an illusion. We can tend and aid the great untamed land, but she can never be controlled. For this I am thankful. Our debates continue, but always, we hope, with the health of the land at the forefront of our minds, listening for scraps of wisdom that can penetrate our technified minds. The oldest and largest oaks we work with and around were youngsters when this land was inhabited by people whose cultures coexisted gracefully with the health of savannas through mutual aid and direct participation. These same oaks witnessed the smothering of those ways. We look to them to guide our actions.

It is often the swooping branches on such trees that come up in long talks around the woodstove, and it is these trees who call us out to work in the day. When an elder oak is opened up, we are restoring dignity to this tree. Often the oldest trees in an area are "fence-line oaks" that were spared from the saw a century ago so the farmer's barbed wire could be tacked up and the border between the wild and domestic made clear. This is a wound I would like to heal but not a scar I want to erase. We need to feel this scar tissue so we might learn. For the wild that remains around and in us, may we learn. ○

Notes

1. Doug Tallamy, *Bringing Nature Home* (London: Timber Press, 2009).
2. Joshua G. Cohen, Michael A. Kost, Dennis A. Albert, and Bradford S. Slauther, *A Field Guide to the Natural Communities of Michigan*, Michigan Natural Features Inventory (Lansing: Michigan State University Press, 2014), 174–178.
3. Tom Wessels, *Reading the Forested Landscape* (Woodstock: The Countryman Press, 1997).
4. Cohen et al.
5. Ahmed Djoghlaf, message delivered on the International Day for Biological Diversity, May 22, 2007.

Jamie Hunyor

CARCASS

an elegy for the unknown sheep

I come upon it searching for the herd among thick bramble,
a makeshift staff in hand, straight, dry, last year's oak.
Picked clean, no meat, no smell,
weeks of rot and decay in the dry heat
covered by the crowded
ammonia aroma of a stagnant herd.
With the toe of my boot
I nudge the skull in profile on the dirt,
turning it to rest on its opposite jawbone,
small piles of loose wool scattered about.

What predator stalks silently
sleeping prey along Wooley Creek?
Heard talk of a mountain lion in these hills,
the creature that must have dragged
Stuart's young wether and buckling
across the rocks of a shallow crossing.
Vultures gathered in the oaks that line
the creek for two weeks,
fighting in the boughs for their turn at the meat.
What kind of shepherd am I,
to have never checked and come upon it later
only by accident?

At night, standing over a shelf full of rock,
I wonder, what stone could mark
this place? The singing curves of
the carcass' ribs are
its own mausoleum,
with blackberry flowers
and wild rose petals strewn around
it, their vegetation creeping hungrily to
cover this unmarked grave.

Kirsten Mowrey

BELONGING IN THE GREAT LAKES

I will readily admit my garden is a bit of a hodgepodge, with some raspberry and blueberry bushes—legacy of a permaculture course I took—some herbs, and a scruffy area in the southwest corner holding dogwood and grapes that run wild, feeding the birds mostly. There's a lawn and beds, but no plan, no cohesiveness: just plants that I bought on sale, either with intention or because I found them pretty, but no vision, no center, no whole. Over the twenty-plus years I have lived here, the garden has been a refuge, a creative space, a chore, an onus, a chance to connect to the natural world, a social change platform, and a hideaway. Currently, I believe it represents my struggle to ground, be connected, and belong.

If mainstream culture says *go go go*, then maybe the answer is to *stay stay stay*. To root here, create traditions, and nurture an energetic, creative, imaginative spiritual relationship with the Great Lakes ecosystem. One aspect of that is through ritual. In *The Wild Edge of Sorrow*, psychologist Francis Weller says we use ritual to "suture the wounds in the broken terrain of belonging." Ritual is not ceremony. While both reinforce community and our connections, ceremony doesn't change us, whereas ritual does. Ritual offers the capacity to alter our souls and selves, to rearrange, repair, and connect with the transcendent.

In *Timaeus*, Plato wrote about the world soul, the *anima mundi*: "a single visible living entity containing all other living entities, which by their nature are all related." This European belief views the world as ensouled,

alive—just as many Indigenous North Americans do. Practices to connect with this world soul are found in remnants of manuscripts and in tales of the folk, those who lived on the land. These stories were originally passed down orally, to guide future generations through pain, loss, and recovery. In one story, a young woman is betrayed by her father, who chops off her hands. She leaves her family and finds a garden with fruit trees. She asks the trees for food, "enough to satisfy her hunger, but no more." She meets and marries a king, is betrayed again, and returns to the forest. Once again she asks for assistance from the natural world, finding a safe harbor where she remains for seven years while her hands grow back.

Here are lessons: Ask the land for help, only take what you need, have patience. All of these lessons are counter to modern Western sensibilities, yet are found in Indigenous stories. We can connect to the world as alive, with human beings as one among many, not one above others.

My maternal great-grandmother was Greek and never learned English. When I was a child, we would always smile at each other and speak in our respective languages, and I knew that what we were really saying was "I love you." That is what I need to do in my garden. I honor this land with ritual that I have created, received, or imagined, alone or with others. My ritual doesn't have the venerability of age, but it does have the authenticity of being mine. I offer verbal thanks when I harvest my vegetables; I offer thanks with food I've made,

Henri-Edmond Cross *The Artist's Garden at Saint-Clair*

or art I've drawn, which I bury in the ground. Honoring the trees and thanking them for shade in the summer, planting sculptures in their roots. Cupping the water in my hands every time I take a shower and singing thanks to the St. Clair River and co-creating ritual with others to show gratitude to Lake Erie, to which the water returns it. Truly settling, inhabiting, and rooting into the world around me, not merely passing through, transiting on my way to something else.

Many of us don't have religion, yet still have needs of the spirit, a way to connect to the sacred, a way to honor the processes of life and death, all the vicissitudes of living. Intentional ritual and giving thanks in expressive arts holds the power to change our hearts, bodies, and culture, to transform our way of walking in the world.

If there is a grand land plan for my Great Lakes garden, it is to find balance and, hopefully, harmony. For European roses to sing next to native sunflowers feeding goldfinch, to feed myself blueberries and cellularly be more of glacial moraine, deciduous forest, grass-bordered lakes. A niche where I can find traction to root, grow, thrive, and belong.   ○

The Tree Economy

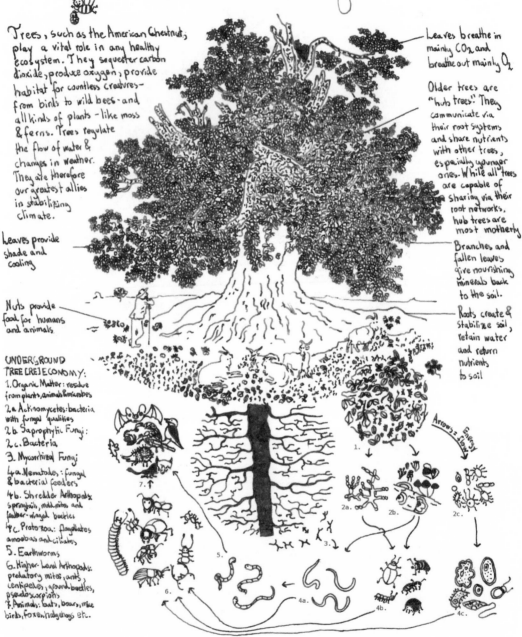

Trees, such as the American Chestnut, play a vital role in any healthy ecosystem. They sequester carbon dioxide, produce oxygen, provide habitat for countless creatures — from birds to wild bees - and all kinds of plants — like moss & ferns. Trees regulate the flow of water & changes in weather. They are therefore our greatest allies in stabilizing climate.

Leaves provide shade and cooling

Nuts provide food for humans and animals.

UNDERGROUND TREE [RE] ECONOMY:
1. Organic Matter: residue from plants, animals & microbes
2a. Actinomycetes: bacteria with fungal qualities
2b. Saprophytic Fungi:
2c. Bacteria
3. Mycorrhizal Fungi
4a. Nematodes: fungal & bacterial feeders
4b. Shredder Arthropods: springtails, mold mites and feather-winged beetles
4c. Protozoa: flagellates, amoebas and ciliates
5. Earthworms
6. Higher-Level Arthropods: predatory mites, ants, centipedes, ground beetles, pseudoscorpions
7. Animals: bats, bears, mice, birds, foxes, hedgehogs etc..

Leaves breathe in mainly CO_2 and breathe out mainly O_2

Older trees are "hub trees." They communicate via their root systems and share nutrients with other trees, especially younger ones. While all trees are capable of sharing via their root networks, hub trees are most motherly

Branches and fallen leaves give nourishing minerals back to the soil.

Roots create & stabilize soil, retain water and return nutrients to soil

Arrows = Energy flow

Hans Kern *The Tree Economy*

Bulletin No. 6

1905

HOW TO BUILD UP WORN OUT SOILS

Tuskegee Normal and Industrial Institute

EXPERIMENT STATION

Tuskegee Institute, Alabama

Geo. W. Carver

TUSKEGEE INSTITUTE STEAM PRINT

In Bulletin No. 6, George Washington Carver writes to poor tenant farmers about how to repair cotton-devastated soils. He saw worn soils as tired but very much alive, observing that decaying "rubbish" of any kind could make rich soil. His field experiments were grounded in the desire to surface the land's latent fertility, a seasonal process available to any farmer with a one-horse plow. Cover cropping, deep plowing, fertilization, and crop rotation were center to his reparative approach. Two key takeaways: "Peanuts should be grown by every farmer," and always use whatever you have. — Makshya Tolbert

Catherine E. Bennett

COMPOSTING ELISE STEFANIK

She's a Republican Congresswoman being groomed for the presidency, her party says—a woman to replace Trump, but not his policies. Over the past six months, I've been mourning the trees killed for her mail-based publicity. The League of Conservation Voters gives her a lifetime grade of 37 percent; her legislative record includes votes against pro-environment funding and bans on offshore drilling.[1] As I stare at her smiling advertisement—"Independent Bipartisan Voice for the North Country"—all I can think of are the toxic inks and how the soil organisms will react.

The glossy page sits atop a budding compost ecosystem—after a hard winter, I stopped counting how many lambs wound up here, piled under clods of manure and a swath of reed canary grass. Now reaching temperatures that cause me to yank away my hand, this beautiful heap is producing bucketloads of rich, crumbly soil from wool and bone. Yesterday, our intern remarked on how amazing it is that the smooth, small perfection of a rib cage could grow in the uterus of a ewe. A month ago, this woman struggled being around insects and manure. She now peers with interest and wonder at the white skull separating from flesh and hair.

Over the past few years, I've been reading about bacteria who consume petrol, and of the fungi thriving on radiation in the Chernobyl Nuclear Power Plant.[2] Cousins to the microbes inhabiting farm humus, these abundant organisms consume chemical compounds and excrete healthy elements. Simply by living, they and my compost companions clean up the toxic chemicals—formaldehyde, xenoestrogen-laden plastics, bisphenol A-filled receipts—that industrialized societies have created.[3]

Research done by the Environmental Working Group warns of the dioxins, chlorine, and heavy metals found in the majority of today's paper products,[4] but recycling isn't an option in my county. Where it does happen, the process for reusing crumpled envelopes and discarded TV boxes involves a water bath. The aforementioned carcinogens are either concentrated into "new" paper products or washed into rivers and soils, settling thousands of miles from where they began. This ineffective system means pollutants never meant to touch human skin are spread further yet, streaming from our kitchen taps[5] and carried in our 100 percent post-consumer waste toilet paper.

In an attempt to alleviate such horrors, I am experimenting, knowing that the thermophilic bacteria in the compost pile will recognize the volatility of paper-pulp chemicals. Composting has been proven to disintegrate molecular bonds and mineralize contaminants in ways unachievable by industrialized methods.[6] Detritivores have the ability to heal, and to renew the world.

Not so long ago, "waste" referred only to that which was biodegradable. There was no other kind of trash, garbage, rubbish, or junk. That which was left behind was food scraps, clay shards, animal skins, and woven plant fibers.[7] Antibiotics and pesticides did not swim through our veins; mercury and lead were not omnipresent in the average household. In 2001, American journalist Bill Moyer had his blood and urine tested. Eighty-four health hazards—endocrine disruptors, asthma-inducing chemicals, PCBs, phthalates, and more—had accumulated in his skin, his bones, his lungs, his brain.[8]

When members of this culture die, will our human bodies feed the soils, or poison them?

Reflecting on this culture that not only crafts life-annihilating substances, but embraces them, my daily existence is mission-oriented:

I've stopped questioning my role as a regenerative agriculturist. For so long I thought to myself, "Other people do much more for the planet. I should be blowing up smelting plants. Tracking threatened woodpeckers. Blocking clearcuts. Not farming." And someday, I will join tribes fighting oil in the Ecuadorian rainforest, sabotage Kansas-based vivisection labs, and protect endangered Sumatran rhinos in Indonesia. But right here, right now, on 114 acres of stolen Mohawk territory in northern New York, I restore Ashworths, a potato variety developed on this land seventy years ago.

If this is my task—to rejuvenate soils, reestablish rare foods, and teach others to do the same—then who is to say it's not just as important a fight?

American Burying Beetles—beetles who lay their eggs in the flesh of dead mammals—have become nearly extinct. But if I were a burying beetle, I'd want my children raised on songs of revolution. As my larvae nibbled on the decaying body of a mouse, I would recite stories of how the destroyers had succumbed. I would breathe in the rich nutrients of somebody else's carcass as I sang to my babies a respect for life. I'd tell them how the colonizers could not stand our presence, our ability to alter the world with

our dreams, our ferocity, and our births, so they eventually faded away. I would teach them how to make food from what was remaining, and as a result, the world would keep spinning.

Because that's what it will take. Elise Stefanik is part of a materialistic system fixated on exterminating the fire of this diverse, planetary ecosystem. No matter how many bulldozers we gut, dams we obliterate, pesticides we ban, or

Jo Zimny *Thiers' lepidella*

liberals we elect, the dangerous products left from the fall of colonial capitalism will have to be dealt with. Sure, we can shutter our paint factories, but who will take care of the hexavalent chromium? If plastic bags are illegalized, is anyone out there to decompose hydrocarbons? And what if this culture continues? If people like Stefanik continue to set the rules of the game? If waterways and soils and airways and livers and hearts are continuously exposed to a flood of toxins?

The world will *not* keep spinning.

Thousands of years from now, our vault-bound skeletons may be excavated by anthropologists, but the burying beetles will have starved. Will there be, then, anyone left to clean up the mess? Will someone—a regenerative agriculturist, a little girl singing "Old McDonald," a family who wants to eat—care enough to start a pile, an ecosystem; to make friends with fungi and partner with protozoa? After industry is gone, the only way to effec-

tively eradicate the contamination is through bio-and mycoremediation. We must rebuild the bridges between ourselves and every other organism.

As I lean grimly against the fence around the compost heap, I see a microcosm of the revolution that has already begun. Stefanik is not the only congressperson whose image is now being digested by saprobes, and she will certainly not be the last. I'd be nervous about the ecological consequences of composting an actual politician, but this? This is a damn good start. ○

Notes

1. "National Environmental Scorecard–Elise Stefanik," League of Conservation Voters, 2019.
2. Heidi Ledford, "Hungry fungi chomp on radiation," Nature News, May 23, 2007.
3. Renee Alexander, "Mushrooms Clean Up Toxic Mess, Including Plastic. So Why Aren't They Used More?", Yes Magazine, March 5, 2019.
4. "Dioxin," Environmental Working Group, July 13, 2010.
5. "2,3,7,8–TCDD (Dioxin)," Environmental Working Group's Tap Water Database," 2015–2017.
6. Chris Rhodes, "Mycoremediation (Bioremediation with Fungi) – Growing Mushrooms To Clean The Earth," Energy Balance, June 15, 2014.
7. Derrick Jensen, Aric McBay, What We Leave Behind (New York, Seven Stories Press, 2009).
8. "Bill Moyers' Test Results," Trade Secrets: A Moyers Report, PBS.

Danielle Stevenson

THE ESSENTIAL INVISIBLE

Fungi in Soil Remediation

> It is only with the heart that one can see rightly; what is
> essential is invisible to the eye.
>
> —Antoine de Saint-Exupéry, *The Little Prince*

When I was a teenager, I went on walks from my family's old farmhouse in the county to the nearest city. I walked along the old railway track through agricultural fields to the industrial cityscape. Sometimes I closed my eyes and imagined this place before settlers drained the marshes for agriculture and industry, its vast wetlands abundant with food and habitat for many interconnected lives. Opening my eyes, I saw expanses of greenhouses whose light blocks out the stars at night, like an Anthropocene aurora borealis, and whose runoff feeds algal blooms which choke fish in Lake Erie. I saw abandoned factories and giant forges, leaking tanks with skulls and crossbones on them. I saw how industrial pollution impacts food systems and how industrial food systems create pollution. There are high cancer rates in the area,[1] and though people didn't like to talk about it, the connection between environmental pollution, food systems, and human health was—and is—clear.

I also saw how the earth addresses pollution and heals itself: sumac and dandelion grow up through the concrete, poplar trees grow through old cars. I saw shaggy mane mushrooms busting through asphalt in a ring-around-the-oil-tank.

This is how I became curious about natural or ecological remediation, or action that remedies degraded and polluted soils by supporting the natural abilities of plants (phytoremediation), microbes (bioremediation), and fungi (mycoremediation) to extract, immobilize, and/or degrade contaminants. The goal of ecoremediation projects is to reduce concentrations of pollutants to safe levels for human and ecological health and to regenerate healthy soil or water. These efforts are needed on contaminated industrial sites as well as parks and sites of food production and harvest. Pollution impacts human health significantly through food systems, farms, orchards, and sites of harvest, making these important places to focus remediation efforts. Farmers, growers, and land managers who've grown plants and made compost already have many of the skills needed to remediate soils. For the unfamiliar, these

Liz Brindley *Soil Life in One Place*

skills are easy to learn and relatively cheap to implement.

A good starting place for any remediation project is to learn about the site and what types and concentrations of contaminants are present. Farm and orchard soils—urban and rural, big and small, old and new—are likely to contain some metals and organic contaminants. Metals such as lead, copper, cadmium, and arsenic are deposited through past and present fertilizer, fungicide and biosolid application, wastewater irrigation, and industrial emissions, and they accumulate because they do not break down. Organic contaminants common on food sites include pesticide and herbicide residues, which can persist for a long time in soils. Though these contaminants are largely invisible, they diminish diversity among living organisms on polluted sites and impact the health of exposed humans and other animals.

It is essential to acknowledge this invisible toxicity so we can truly resolve it. With the out-of-site/sight, out-of-mind approach of conventional remediation, contaminated soils are dug and dumped offsite, displacing the problem elsewhere and disproportionately impacting people of color, Indigenous communities, and the working class. We must more thoughtfully consider the methodologies we use.

Before deciding on a remediation plan for a site, observe the life already present. What plants, microbes, fungi, and other organisms are present onsite as a volunteer cleanup crew? Look deeply. Fungi play essential roles in soil ecosystems, though many are invisible to the human eye and have long been overlooked. Underground, there may be hidden helpers connected to the roots of plants, where you'll hope to find arbuscular mycorrhizal fungi (AMF), which associate with 80 percent of all plants. These symbiotic fungi trade nutrients and water for photosynthetic products from the plant and help their partners survive harsh conditions and contaminated soils. This is especially beneficial in phytoremediation and ecological restoration efforts, where AMF inoculation increases plant survival and remediation success. AMF also protects crops (and humans) by inhibiting metal uptake into edible parts, a big help given the low levels of metals common in agricultural soils. Decomposer fungi—notably oyster, turkey tail, shiitake, and button mushrooms—have shown potential to degrade organic contaminants from diesel and asphalt to dichlorodiphenyltrichloroethane (DDT) and polychlorinated biphenyls (PCBs) through their digestive enzymes. Decomposer

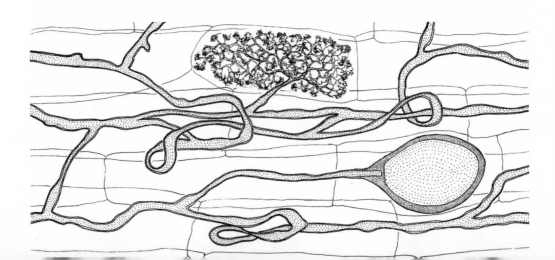

fungi initiate processes of degradation, but the soil is ultimately healed through the collective action of many fungi, bacteria, and plants. More trials with fungi in the real-world settings where contaminants occur would help us better understand the contexts they are best suited for and protocols for successful application. Keep in mind that there are likely indigenous fungi and bacteria already on the contaminated site better suited to the site conditions and capable of transforming the contamination present. See, for example, the story of the symbiotic fungus helping dandelions survive while degrading pure oil tailings in the Canadian oil sands, discovered by researchers from the University of Saskatchewan.[2]

After the agrochemical application stops, after the industrial activity on any site stops, contamination may remain, but life has a chance to grow again. As ecological designer Nance Klehm put it: "Life replenishes itself, soil cleans itself, bodies heal themselves."[3] Birds drop seeds, squirrels plant trees, winds carry seeds, spores, and microbes, and fungi weave their mycelium through the soil from neighboring natural areas. Ants carry mycorrhizal spores which sprout in the presence of plants, helping them establish in degraded soils. Plants and trees like poplars and willow take up metals into their heartwood, locking them up for decades, and microbes around their roots degrade polycyclic aromatic hydrocarbons (PAHs), widespread pollutants. Early successional plants we call weeds, such as mallow, thistle, nettle, dandelion, chicory, asters, buckwheat, and brassicas, are able to grow in poor soils, contributing to soil health by creating shade and organic matter as they drop their leaves in annual rhythms. Some plants act as hyperaccumulators, drawing metals out of the soil into their above-ground parts which can

then be harvested and contained as a cleanup strategy.

Decomposers play important roles in building healthy soil. Earthworms and pill bugs, soil fungi, molds, yeasts, and bacteria- and nitrogen-fixing rhizobia all degrade organic contaminants and immobilize metals just by going about their life processes. These small, hidden organisms are essential to soil remediation. Choose approaches that amplify their life cycles by ensuring suitable water, air, temperatures, and easy food sources like organic matter and compost on the site. While some highly contaminated sites may benefit from reintroduction of beneficial microbes, fungi, and/or plants in thoughtful interventions, consider the possible consequences of these introductions. Create a long-term plan for containment of any toxic by-product of your remediation, such as metal-laden plants used in phytoextraction.

To engage in remediation and regenerate soils, we need to look to the fungi, bacteria, and plants who re-turn pollutants into soil: they are showing us the way. ○

Notes

1. Michael Gilbertson and James Brophy, "Community Health Profile of Windsor, Ontario, Canada: Anatomy of a Great Lakes Area of Concern," *Environmental Health Perspectives* 109, (2001): 82–843. See also "'Strikingly high' cancer rates found in 5 Ontario industrial cities," *Canadian Occupational Safety*, August 26, 2019.
2. T.S. Repas, D.M. Gillis, Z. Boubakir, X. Bao, G.J. Samuels, S.G.W. Kaminsky, "Growing plants on oily, nutrient-poor soil using a native symbiotic fungus," *PLoS ONE* 12, no. 10 (October 2017).
3. Nance Klehm, Jacob Blecher, Martin Brown, Evon Izquierdo, Victoria Thurmond, *The Ground Rules: A Manual to Connect Soil and Soul* (Illinois: Half Letter Press, 2016).

Alice Blackwood

RECKONING

They don't tell you this
at the tourist information centre,
but this bush-razed plain
was once the blue-print for inland invasion;
slowly stripped of topsoil
in a thirsty search for inert shining.

Somewhere beneath
these cankerous suburbs
there's a sedimentary memory
of quolls and bandicoots,
chains of ponds,
bloodshed and rebellion.
Or maybe it's gone—
washed down the river
in the big floods a century ago
and we're stubbornly still
trying to build our homes
on anaemic subsoil;
that might explain the silence.

It is easy to say that
no poems live here
or that they are as sparse
as the kurrajong trees,
those paddock sentinels
casting gaunt shadows
over diminished herds.

It is easy to want to abandon
its fluorescence
for loud honeyeater mountains.

Yet, here I am still,
digging a hole,
planting a plum tree.

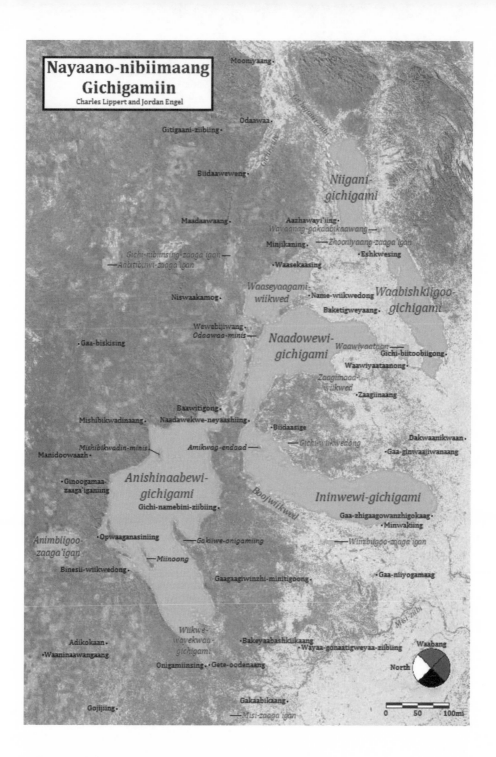

Nayaano-nibiimaang
Gichigamiin
Charles Lippert and Jordan Engel

↑ **Don Davis for NASA**   *One of my earliest* Space Colony *paintings was based on the giant 'Model 3' cylindrical habitats envisioned by Gerard O'Neill. I imagined the clouds forming at an 'altitude' around the rotation axis. At this time the scene is bathed in the ruddy light of all the sunrises and sunsets on Earth at that moment as the colony briefly enters the earth's shadow, out at the L5 Lagrangian point where stable locations are easily maintained.*
→ **Michael Kotutwa Johnson**   *Hopi Corn Colored*

Sky • Vision • Seeing

Air Board, NASA

Adam Calo

WHAT SCOTLAND CAN TEACH BEGINNING FARMERS ABOUT LAND REFORM

Nick Townsend *Ulva Isle, Scotland*

They think about land differently in Scotland: what it's for, what it should be for, who gets to own it, and in all these cases, how to decide. That's why I came here, to learn more about a land reform agenda that pervades both community activists and the halls of the Scottish Parliament.

For a researcher who studies farmland access, the Scottish context provides something unique. Here is a country in the so-called Global North with a strong historical commitment to private property and cultural comfort with landowning "estates" that, nevertheless, is now drawing a line in the sand with regards to land consolidation.

The changes, written into a suite of land reform acts passed between 2003 and 2016, free the country's land from the dominance of private land ownership and establish a diversity of alternative tenure arrangements, promoting the goal of reaching one hundred million acres of community land ownership by 2020.[1] A parliamentary report in 2013 found that land equity is still incredibly lopsided, as an estimated four hundred thirty-two individuals own 50 percent of the private rural land holdings.[2] This land reform ambition could be considered akin to correcting the course of a massive trawler, but the progress thus far is stunning.

As an American who would be escorted out of any congressperson's office for saying the words "land reform," my jaw frequently drops.

Consider the members of the North Harris Trust, for example. In 2003, they purchased a 22,500-acre estate, bringing the land into community control when the landowner (a cider magnate from England) put it on the market. When the entire five thousand-acre Isle of Ulva was put up for sale, residents there organized and invoked their newfound rights to community ownership, eventually succeeding. The story of Action Porty, a community group in Edinburgh, is equally inspiring: they were able to use the land reform legislation to prevent the sale of their local church for development and plan to transform it into a community hub.

As an American brought up on the dream of property ownership, what has always been the most remarkable about places going from "up for sale" to "owned by the community" is the role of the Scottish Government in facilitating these asset transfers. Land reform is part of the government's legislative platform, spoken freely as rational policy. It is based on the idea that moving land from large, privately owned interests into community ownership will increase productivity, rural well-being, and measures of sustainability and resilience.

The first set of the Land Reform Acts[3] grants properly formed community bodies (a group representing a geographic area with compliant legal status) the first chance to buy land in the event of the sale. This is what's known as "right to buy," a pre-emptive right granted to communities who register their interest in the asset in a national register.[4] A second wave of legislation gives communities the right to compel landowners to sell if the land was deemed to be vacant or in neglect. Through a dedicated land fund, communities can bid for up to one million pounds of public funds to carry out purchases if land is passing into ownership by the community entity, often taking the form of a benefit society or local trust. In some cases, the Scottish Land Fund provides greater financial support, like the £4.4 million granted to the community group that purchased the Isle of Ulva.

Recently, the most exciting, yet untested, piece of Scottish land reform has come into force, known as the Right to Buy Land to Further Sustainable Development. In this provision, passed in 2016 but come into effect in 2020, assets that are deemed necessary for the "sustainability" (often perceived as the longevity of community existence) of a community are subject to compulsory sale if a strong enough development plan is put forward.

The measure, which is now the law in Scotland, is utterly radical. If a group successfully argues that any asset would better serve the goals of sustainability if it were managed under the community's control, the current landowner must sell the land. Wherever you are, think about a piece of farmland, or a building nearby that could be utilized to further the sustainability of your community. Imagine what having the power to own that asset could do to change the landscape of American agriculture!

But who decides what is sustainable? The decision not to clearly define sustainability is either the law's failure or its genius. Governmental advocates have repeatedly stated that the hope of the Land Reform Acts is to compel a new culture of fair land use negotiations rather than repeatedly exercise compulsory sales. The prospect of having a community rise up and force an estate owner to sell should motivate new forms of negotiations between those who own the land and those who live on it. Thus, while the Land Reform Acts grant new entitlements, they also rebalance the social power to make decisions about land.

In Scotland, the message of land reform is that if you hold the title to land, you also have the responsibility to ensure that its use is balanced in a way that promotes private profit and public good. Thus, while knee-jerk reactions can conjure images of forced redistribution, land reform here may really be about changing the way a nation thinks about property ownership and the public good. The public-good dynamics of agricultural land have long been marginalized by the logic and legal normalcy of individual land ownership. In places where strong ownership models persist, alternative visions of agricultural land use for the public good will consistently be frustrated.

For most farmers' movements of the United States, the Scottish approach to land access appears wholly alien. The dominant approach to land access in the US is through the logic of private property. Individuals go to heroic lengths to bring their goods to market and sell a business plan to a sympathetic investor. Only then can a farmer raise the capital to purchase land and persist, happily ever after. In many beginning farmer narratives, land ownership is where the story ends.

But I think it's the wrong story to tell.

When a beginning farmer can buy property in the US, they usually end up buying up one of the marginal parcels that remains after being picked over by agribusiness. An individual landowner is often saddled with the debt of the purchase, which may restrict their ability to farm how they want or to invest in alternative farming activities not directly aligned with capital-intensive agriculture.

The ideal of individual land ownership pits beginning farmers against each other in a competitive land market where the deck is already stacked. By instead working collectively, beginning farmers can rethink what land is and how it should be used. Rewriting the way we access, own, and transfer farmland is the key ingredient for a social movement determined to provoke a just and transformative food systems change.

I don't imagine that the federal government in the US will soon put forth a land reform agenda of its own. But that doesn't mean the lessons from Scotland can't be applied to create a paradigm shift in the US beginning farmer movement. Farmers could demand "right to buy" provisions in their leases and regional governments or farmer organizations could incentivize the same. Where farmland is controlled by a few interests or individuals, beginning farmers could

Harper Lake, California

demonstrate how control of some of the land leads to improvements in the environment, food production, and community life.

In the current ecological and political moment, beginning farmers face an enormous burden. They are tasked with repairing hollowed out rural areas and creating a restorative food system from scratch while reversing trends of climate change. In exchange for doing this labor, they should also demand a mechanism for gaining control of the land. For inspiration, the rise of community land ownership in Scotland provides a radical template. ○

Notes

1. The last governmental accounting puts the figure at 518,451 acres of community-owned land. Source: Rural & Environmental Science and Analytical Services, "Community Ownership in Scotland: 2018," December 11, 2019.

2. James Hunter, Peter Peacock, Andy Wightman, and Michael Foxley, "432:50 – Towards a comprehensive land reform agenda for Scotland," Scottish Affairs Committee, July 11, 2013.

3. The "set of Land Reform Acts" refers to a series of acts and statutory interventions put into law between 1997 and 2016 that provide new pathways for community ownership of land and other assets. The Transfer of Crofting Estates (Scotland) Act 1997 was an act of the UK Parliament that gave community bodies a mechanism to acquire particular forms of agricultural lands in the Highlands and Islands of Scotland (known as crofts) in cases where the state was previously the owner. The 1997 act established the idea of "community" as owner and influenced the measures enacted by the Scottish Parliament in the years that followed. The Land Reform (Scotland) Act 2003, in addition to establishing public-access rights for purposes of travel and recreation—the so-called "right to roam"—introduced the power of first-right-of-refusal for community bodies in rural Scotland and gave crofting communities a right to force a sale from a private owner. The Community Empowerment (Scotland) Act 2015 extended the right-to-buy into urban areas, built a mechanism for community asset transfers from public sector entities to communities, and introduced a power for community bodies to exercise a right-to-buy in cases without a willing seller, or if the land was recognized as abandoned, neglected, or used in a way that was environmentally detrimental. Finally, the Land Reform (Scotland) Act 2016 introduced a right- to-buy for sustainable development, which allows for a forced transfer of land to a community when the existing land use is acting as a block on sustainable development, and the transfer not taking place is likely to cause harm to the community.

4. The community bodies must first successfully register their interest in acquiring the asset in a public register, the Register of Community Interests in Land (online at *rcil.ros.gov.uk/RCIL/default.asp?category=rcil&service=home*), and the plan must be approved by appointed ministers.

Rhamis Kent

A UNIFIED THEORY OF EARTH REPAIR

Is Fixing Trophobiosis the Key to Beating Everything from Coronavirus to Locust Swarms to Climate Change?

The problems of land degradation and desertification have been linked to chronic drought and flooding, along with the associated soil erosion and reductions of soil quality, including organic matter and nutrient content. Many parts of the world frequently experiencing these events are also prone to catastrophic and highly destructive infestations of pests (such as locusts) and increased incidence of disease in both vegetation and humans alike. Perhaps more relevant now, COVID-19 did not emerge from a vacuum. The novel coronavirus, like other viruses that have made an appearance in recent years (SARS, H1N1, HIV), is zoonotic in origin and can be attributed to loss of crucially needed animal habitat that serves as a buffer capable of limiting contact between animals and humans.

These breaches of environmental and ecological boundaries have, again, been translated directly into immense economic and social disruption as well as increasingly precarious situations with regard to human security and stability. Yet there is another underlying condition to be included with this assessment, which is imperative to fully comprehending the best solutions. It's a phenomenon known as trophobiosis.

Trophobiosis is a word derived from two Greek roots: *trophikos* (nourishment) and *biosis* (life). As French agronomist Francis Chaboussou writes, "the relationships between plant and parasite are primarily nutritional." According to the trophobiosis theory Chaboussou presents in *Healthy Crops: A New Agricultural Revolution*, it is nutrient deficiencies and imbalances that lead to pest and disease outbreaks, and synthetic pesticides and fertilizers that cause such deficiencies and imbalances.

Trophobiosis is a symptom or product of land degradation and poor land management. Issues such as the periodic occurrence of locust infestations and the prevalence of other pest and disease attacks can be seen as manifestations or expressions of trophobiosis.

This condition is exacerbated by excess concentrations of carbon and other greenhouse gases in the atmosphere—a problem largely attributed to long-term changes in land use that involve the removal of Earth's major terrestrial carbon storages, coupled with widespread reliance on atmosphere-polluting technologies—which have been found to impede the ability of vegetation to take up essential nutrients. The overall effect can be observed in alterations made to global climate, food quality, and ecosystem function, all for the worse.

The following citations provide an overview of foundational components forming the development of a "unified theory" of earth repair. The aim is to facilitate humanity's ability to solve several urgently pressing contemporary challenges simultaneously through well-informed and well-directed practical action.

"Pests shun healthy plants. Pesticides weaken plants. Weakened plants open the door to pests and disease. Hence pesticides precipitate pest attack and disease susceptibility, and thus they induce a cycle of further pesticide use," writes environmental scientist John Paull.[2]

According to Chaboussou, "We need to overcome the idea of 'a battle'; that is—we must not try to annihilate the parasite with toxins that have been shown to have harmful effects on the plant, yielding the opposite effect to the one desired. We need, instead, to stimulate resistance by dissuading the parasite from attacking. This implies a revolution in attitude, followed by a complete change in the nature of research."[3]

In taking a non-military approach to farming, Chaboussou is taking a position in solidarity with early proponents of organic agriculture including Rudolf Steiner, Ehrenfried Pfeiffer, and Lord Northbourne, who coined the phrase "organic farming" in 1940.

Pfeiffer puts it this way: "When the biological balance is upset, degeneration follows; pests and diseases make their appearance. Nature herself liquidates weaklings. Pests are therefore to be regarded as nature's warning that … the balance [has been] sinned against."[4]

Northbourne writes, "There can be no quarrel between ourselves and nature any more than there can be between a man's head and his feet…. We have invented or imagined a fight between ourselves and nature; so of course, the whole of nature—which includes ourselves as well as the soil—suffers…. We have tried to conquer nature by force and by

intellect. It is now for us to try the way of love."[5]

From the outset of organic agriculture, Northbourne understood the escalation problem of chemical farming. "True, plants continue to grow on the ground, but why is it that more and more spraying is necessary?" he writes. "What answer can there be but that diseases and pests are more and more rampant; in spite of the fact that far more is being done, far more skill is being lavished on the combating of disease than was ever dreamed of in the past…. It is liability to disease and not disease itself which indicates ill-health."[6] This is the issue that Chaboussou addresses with trophobiosis theory.

The theory has been summed up by ecologist and molecular biologist Dr. Ulrich Loening as follows: "Most pest and disease organisms depend for their growth on free amino acids, and reducing sugars in solution in the plant's cell sap. Every farmer has experienced the increase in diseases after heavy [fertilization] with nitrogen; the Green Revolution varieties are good examples in which rich [fertilization] creates susceptibility to pests, requiring more pesticides to control. Chaboussou explains why. Almost all conventional chemical agricultural technologies create favorable conditions for the growth of pest and disease organisms…. The susceptibility of the crop is increased: when offered free nutrients, pests grow better and multiply faster. In this sense therefore, agro-chemicals and poisons cause pests and diseases."[7]

Under Chaboussou's theory, an excess within the plant of less complex biochemical molecules, such as amino acids (rather than proteins that they build) and/or simpler sugars such as glucose (rather than the more complex carbohydrates such as glucose polymers and other polysaccharides) offers an attractive milieu for pests and disease.

Resistance and susceptibility to attack are a function of the nutritional state of a plant. When proteins are being synthesized, the plant is resistant; when proteins are being broken down, the plant is at risk. "Organophosphates inhibit protein synthesis," Chaboussou writes. "This is the cause of the plant's increased susceptibility, not only to sucking insects such as mites, aphids, aleurodes [aphids] and (so it seems) psyllids but also to diseases, fungal and otherwise."[8]

Chaboussou reports "a parallel between the effects of herbicides and those of nitrogen [fertilizers]," explaining that "the pesticides that contain nitrogen—practically all chemical pesticides—are cations [positively charged ions]. They can replace cations such as Ca, Mg, and Zn from the exchange complex."[9] Hence, applications of herbicides and synthetic fertilizers can lead to deficiencies in the treated plant.

Chaboussou states that "artificial organo-chemicals have a very special affinity for plant tissues" and pesticides applied to the leaves (foliar application) find their way into the body of the plant.[10] This can be through the cuticle and through the stomata. Since light promotes the maximum opening of the stomata, penetration of pesticides will be greater where the poison is applied in daylight. Penetration of pesticides into the body of the plant can be via the leaves and also the roots, seeds, and branches. Having penetrated the plant, pesticides can be transported through it via both apoplastic (extracellular) pathways and symplastic (intracellular and intercellular, or within a cell and from cell to cell) pathways.[11] The plant, so weakened, is susceptible to pests and disease.

The authors of "Living With Locusts: Connecting Soil Nitrogen, Locust Outbreaks, Livelihoods, and Livestock Markets" suggest that similar imbalances result from overgrazing and soil degradation. Locust outbreaks, for instance, have been correlated with plant quality, which depends on the nutrients that plants can extract from soils.[12] Soil quality, of course, is influenced by land-use practices, especially on farms and rangelands. In Mongolia, research has tied locust outbreaks to the depleted nitrogen content in plants found in overgrazed, eroded pastures.[13] As noted above, a number of studies[14] have discovered that increased levels of atmospheric carbon coupled with a warming climate further restrict the ability of plants to absorb nutrients, making them not only less healthy, but also more susceptible to pest attack.

How are these nutritional imbalances best addressed? Focusing efforts on improving soil quality and biological activity through intelligent land management is clearly the path that must be taken. There is no other way.

Engraving by J.F. Castro   *The Plague of Locusts*

Soil biology (in all of its forms) must be permitted—and helped—to determine the chemistry best suited to maintain the functional health of landscapes globally. The primary purpose of earth repair is to create the habitat and conditions conducive for these biological elements to perform the work they're built for doing—and to refrain from undermining their ability to do it! ○

This essay was adapted from a blog of the same title, originally published by Worldwide Permaculture Association.

Notes

1. Francis Chaboussou, *Healthy Plants, A New Agricultural Revolution* (Charlbury, United Kingdom: John Carpenter Publishing, 2004).

2. John Paull, "Trophobiosis Theory: A Pest Starves on a Healthy Plant, " *Journal of Biodynamic Agriculture Australia*, no. 76 (January 2007).

3. Chaboussou, 209.

4. Ehrenfried Pfeiffer, *Agriculture Course: The Birth of the Biodynamic Method* (Forest Row, United Kingdom: Rudolf Steiner Press, 2004): 16.

5. Lord Northbourne, *Look to the Land* (London, United Kingdom: Dent Press, 1940), 191–2.

6. Northbourne, 101.

7. Ulrich Loening, in Francis Chaboussou, 2004 trans., *Healthy Plants, A New Agricultural Revolution* (Charlbury, United Kingdom: Jon Carpenter Publishing): x, xi.

8. Chaboussou, 55.

9. Chaboussou, 156.

10. Chaboussou, 39.

11. Paull, 24.

12. Arianne J. Cease, James J. Elser, Eli P. Fenichel, Joleen C. Hadrich, Jon F. Harrison, and Brian E. Robinson, "Living With Locusts: Connecting Soil Nitrogen, Locust Outbreaks, Livelihoods, and Livestock Markets," *Bioscience 65, no. 6 (June 2015).*

13. Cease et al.

14. See "Increased carbon dioxide levels in air restrict plants' ability to absorb nutrients," *Science Daily*, June 12, 2015; "Availability of nitrogen to plants is declining as climate warms," *Science Daily*, October 22, 2018; and Helena Bottemiller Evich, "The Great Nutrient Collapse," *Politico*, September 13, 2017.

M.T. Samuel

lime

last night ariadna dreamt of   my wedding
   kevin in a green suit
green like spruce needles   i wondered
green like a tin roof or maybe green like a lime

which is the size of my sister's baby
i'll call it a baby since it's the size of a lime

  and we are speaking in terms of its growth

on a tuesday of wood stove sending pin pricks up the pipe
unemployment office hold songs and scrap paper math
curtains still closed and room temp coffee still going down
it's hard to imagine a lime size   baby
or a groom   in   green
but i'll take these omens
stir them into my coffee
and plan for a future that looks like that

Amy Franceschini

FREE SOIL: REPORT NO. 2

A Brick Moon and the Farm that Sailed Away

> Sailors have the fortune of being with the stars in their labor.
> Farmers tend to sleep during night hours, but pay close attention
> to the orientation of both astronomical objects and the millions
> of twinkles of life under our feet, we call soil.
>
> — The Farmer and Sailor

1975 – Stanford Torus

During the summer of 1975, a group of engineers, phys-
icists, geologists, social scientists, architects, students, and volunteers constructed a convincing picture
of how people might permanently sustain life in space
on a large scale. The Stanford Torus is a donut-shaped
ring, one mile in diameter, that rotates once per minute
to create Earth-normal gravity and was designed to
house ten thousand people.

In *Space Settlements: A Design Study*, the participants
wrote:[1]

> Humans living in space must have an adequate diet;
> and food must be nutritious, sufficiently abundant,
> and attractive. There must be enough water to sus-
> tain life and to maintain sanitation. A diet
> adequate for a reasonable environmental stress and
> a heavy workload requires about 3000 Cal/day. It
> should consist of 2000 g of water, 470 g dry weight
> of various carbohydrates and fats, 60 to 70 g
> dry weight of proteins, and adequate quantities of
> various minerals and vitamins. The importance
> of the psychological aspects of food should not be
> neglected. The variety and types of food should
> reflect the cultural background and preferences of
> the colonists.

← **Don Davis for NASA** *The Stanford Torus*
The 1975 NASA Ames/Stanford University Summer Study worked out the broad engineering requirements for a toroidal-shaped space colony
design. The challenge of sustaining something like a closed ecosystem was a theme I wanted to emphasize.

2016 – The Misting Miner

Alexey Buldakov of Urban Fauna Lab in Moscow material-
izes the invisible phenomenon of mining cryptocurrency.
The excess heat produced by the machinery in Misting
Miner as it performs this process is a latent and
untapped source of energy. His sculpture harnesses that
energy and reveals its transformative potential by
turning it into fog through the water cooling system
inside it, which he reroutes to follow a cycle of evaporation, condensation, and precipitation.

1977 – Faraway in the Future, A Moisture Farm

> *Harvest is when I need you the most. Only one more*
> *season. This year we'll make enough on the harvest*
> *so I'll be able to hire some more hands.*
>   –Owen Lars, to Luke Skywalker

The Lars homestead was one of several moisture farms
that made up the Great Chott salt flat community, located
on the desert planet, Tatooine. Mostly underground and
accessed through a pourstone entry dome, the homestead
was a warren of interconnected rooms and vast storage
areas. Being a moisture farm, there were multiple moisture vaporators scattered around the property, many of
them decades old, capable of collecting water from the
dry air. Any water obtained was used either for consumption or to water the farm's marginally profitable
hydroponic garden. To conserve power from the property's supercharged bio-converter power generator and
small fusion-cell generators, the compound was shut
down after nightfall.

1869 – The Brick Moon

Everett Edward Hale's serial novella, *The Brick Moon*,
is possibly the first account of a space colony. A brick
sphere intended for guiding maritime navigators is cat-
apulted into Earth's orbit by rotating wheels. When it

NASA *The Brick Moon*

rolls onto the catapult too soon, still containing many
workers inside, the first space station is launched.
Fortunately, the workers have ample food and supplies
(even a few hens), and they decide to live the good life

permanently in space, maintaining contact with Earth only through a Morse code signaled by making small and large jumps from the external surface of their tiny spherical brick colony.[2]

2018 – Regolith + The Effects of Micrometeorites in Extraterrestrial Building Materials

In a small laboratory in Palo Alto, young engineer Mia Allende forms small bricks by hand. She is using a binder for regolith (moon dust) derived from proteins extracted from cow blood to create a "super concrete" that could potentially allow for building structures on the Moon and Mars. She uses the Ames Vertical Gun Range to test her small creations. Each brick is placed on a stage inside a vacuum chamber where high-velocity projectiles are shot at the structure. The force of this 150-foot gun vaporizes Allende's fine handicrafts. A high-speed camera captures the impact and is studied in slow motion to show the fractures in the material that mimic the effects of micrometeorites on building materials.

Allende and others aim to make a self-healing substrate that serves as a shield that will break up the projectile into a cloud of material that expands and is distributed over a wide area around the dwelling.

2019 – Bending the River Back into the City

> We "undevelop" this historic land by lifting the asphalt and letting the soil and seeds underneath reconnect with air, light, and water. My purpose is to demonstrate and support a living system that has been largely sealed up by the strategies of development.
> —Lauren Bon, Artist/Founder, Metabolic Studios, Los Angeles

← **NASA** *Ames Vertical Gun Range*
Dr. William Quaide and Donald Gault of Ames Research Center's planetology branch
used this gun range to study the formation of impact craters on the Moon.

Bending the River liberates soil, water, and seeds that made up the pre-settler colonial floodplain at Yaangna, "the place where the wild river would regularly flood." Using a waterwheel seventy feet in diameter, the project diverts water from the Los Angeles River. A ten- by eighty-foot pit is excavated for the waterwheel, and an inflatable dam impounds water, sending it into a pipe that feeds the waterwheel pit. Water is lifted by the wheel, filtered, disinfected, and used for irrigation.

For one month in the summer of 2019, the concrete jacket of the LA River (created as a flood control measure) was pierced to allow for water diversion and access to flowing water and flora that had not seen light since the channelization of the river began in 1938. A few of the long list of seedlings presenting themselves in the floodplain sediment unearthed that summer:

> Horseweed *(Conyza canadensis)*, brass buttons *(Cotula coronopifolia)*, mustard *(Brassica)*, peppergrass *(Lepidium)*, and smallseed sandmat *(Euphorbia polycarpa)*. ○

Notes:

1. Richard D. Johnson and Charles Holbrow, eds., *Space Settlements: A Design Study,* National Aeronautics And Space Administration, 1977.
2. Everett Edward Hale, *The Brick Moon, Atlantic Monthly,* vol. XXIV (July–December 1869) and vol. XXV (January–June 1870).

Michael Kotutwa Johnson

REFLECTIONS OF A HOPI FARMER

Interior of a Hopi Indian house at Oraibi, Arizona, ca.1900

I am a traditional Hopi dryland farmer and I raise a variety of heritage-based crops including corn, beans, melons, squash, and gourds. I am grateful that my people and family have been farming for at least two hundred fifty generations, which I often say rivals all farmers in the Midwest. Our Hopi land, about ninety miles northeast of Flagstaff, Arizona, only receives about six to ten inches of annual precipitation each year. Hopi is located on the Colorado Plateau, which consists of desert shrubs, grasses, and soils typical to semi-arid

stories tell us we were given a planting stick, a gourd of water, and seeds. We were farmers before we had ceremonies. But I've learned from a Hopi elder, Harold Joseph, that we were told we needed faith to raise crops as we do. These instructions were given to us when we first arrived in our location millennia ago.

As a youth, I spent summers with my grandfather learning various techniques to plant and maintain our crops. Later, after I graduated high school, I went to Cornell University as an undergrad to earn a degree in agriculture. While in college I often wondered why the typical American farmer had to use pesticides, herbicides, and various soil supplements, and why they had to purchase huge equipment. I imagined what a fourteen-row planter would look like going down my family's one- to five-acre fields and finishing in five minutes. (Growing up, it usually would take me and my relatives a couple of days to plant by hand.) An agricultural economics class provided understanding: farmers in America are locked into an economic formula that demands high yields, which are tied directly to efficiency. They simply do not have a way out, and the cost is financially and psychologically devastating.

At Hopi, economics is not the primary reason we farm. We do not farm to make money, but to survive and carry on our ancient ceremonial customs. In so doing, there is no separation between our spirituality and our agriculture system. As a result of those two factors coexisting, we have formed a unique partnership that has withstood several shocks—environmental degradation, non-traditional forms of government—and our way of farming and ceremonies continue to this very day.

Today, academics call what Indigenous people have been practicing since time

Hopi boy planting corn, 1918

environments. A visitor once asked me after looking at our dry region, "How come you guys do not irrigate? You would have a crop every year." I turned and smiled and said, "If we irrigated, then what would we pray for?"

Faith is at the heart of why we have continued to farm in our semi-arid region for over two millennia. When we first arrived here, our

immemorial Traditional Ecological Knowledge (TEK). I call it "the Hopi Way of Life." This knowledge was gained by thousands of years of observation, practice, trial, and error. What was gleaned is a unique knowledge of environment and of heritage Hopi seed varieties that can survive the region's environmental conditions. Despite our more than two thousand years of replication in crop production, our techniques are often called primitive and economically insufficient. Conventional agriculture and, now, regenerative agriculture, only go back about a few hundred years.

We are stewards of the land and our agriculture system has been developed and replicated in the location where we reside. Our crops and agriculture techniques have been adapted to overcome climatic events like drought. All of our agriculture methods are designed to conserve soil moisture. I know of no place else in the country that raises as many original heritage crops using time-tested techniques for agriculture and conservation. At Cornell, agronomists and crop scientists told me I needed thirty-three inches of rainfall a year to raise corn. I thought to myself, *these professors do not know what they are talking about.* They outlined the mechanics of how corn is supposed to be planted, such as placing kernels at one-inch depths, two inches apart, and in fourteen-inch rows. I remember telling them at Hopi, we plant our corn anywhere from six to eighteen inches deep, depending on soil moisture availability, in six-foot rows, with ten to twenty kernels in each hole. The only technology we have is our planting sticks and the biodiversity of our seeds.

I farm this way because this is what I was told by my grandfather would work. He learned from his father, and he his, and on and on. It was not until I attended Cornell and again

while getting my PhD in natural resource studies at the University of Arizona that I was asked to prove that our techniques worked. Having to prove my way of farming was in essence asking me to prove my faith, and at times I was really at the crossroads of whether to continue my education. However, I did and have had numerous opportunities to be heard.

My grandfather explained how to prepare for droughts. He would tell us that we must always plant enough to try to have three to five years of seeds in storage. In so doing, we would not starve. For my grandfather, it was never based on generating profit in the form of commodities, but rather to feed his family. Even in the era of the Great Depression, he said, Hopi families did not go hungry like many in the United States. He said that Hopi people just went about their lives uninterrupted.

When I was building my Hopi house out of sandstone and mud mortar, as we have been for centuries, an elderly Hopi gentleman stopped by for a visit. He saw me plastering my walls and asked if I'd remembered to place some seeds in the plaster mix. I gathered some seeds from my seed room and did what he

Milton Snow *Corn and melons. Sam Shingoitewa's farm. 15 Miles SW of Toreva Day School. Sam Shing [Shingoitewa] working. October 1, 1944.*

suggested. After a week, he stopped by again. He asked if I'd done what he asked me to, and I told him I had, but I wanted to know why— those seeds would not grow. He paused and smiled. Some Hopi used to do that a long time ago because it would remind them, he said, that they always had food in the house, especially when crops were not raised. His lesson was that farming is not only about physical well-being, but also psychological, and ingrained in our spirituality.

Someday, I would like to purchase two or three tractors and start a co-op of Hopi farmers; however, trying to find revenue to begin such a task has proven difficult. Our land is held in trust by the federal government, like most Indian land, and access to capital from loaning entities like the Farm Service Agency (FSA) has long been limited, a result of discriminatory policies and practices.

Being a Hopi farmer is not just about putting seeds in the ground; it is about treating and raising those seeds like they are your children. We sing and talk to them as they grow. We touch those seeds at least seven times from harvest to planting. It is a relationship we have with what we raise that most people do not seem to understand, built by countless years of experience. It must continue. Faith is the underpinning of what it is to be a Hopi farmer, and it is also the recognition of our cultural identity and its value that must be enhanced. For without faith and recognition, our way of life may slowly disappear. To be a Hopi farmer is one of humbleness and quiet reflection; I have become truly grateful for what I do have. Perfecting the process to raise crops in a semi-arid region was long, but it was the journey and shared knowledge of generations of Hopi people that allowed us to survive, and which will carry us forward in the future. ○

Maria Elena Peterson, Courtesy of Michael Kotutwa Johnson   *Many Hopi families teach their children how to raise corn by hand*

Lily Piel

GROUNDBREAKERS

Mainers Shaping Agriculture's Future

Jim and Megan Gerritsen
Talented farmers who have stood up
to Monsanto, the Gerritsens display
in spades the elements of character found
in many Maine farmers—independence,
smarts, and courage.

Caitlin Hunter (with Tiny Turner)
Farmer and artisan cheese-maker, Hunter has been a force behind the Maine Cheese Guild and a mentor to many.

Leah and Marada Cook
The Cook sisters run Crown O'Maine Organic Cooperative, demonstrating daily that out-of-the-box solutions can be delivered on a truck.

Chris and Dave Colson
Among the first farmers in the Northeast to make a living by growing organic produce and selling
it locally, the Colsons have grown food and trained young farmers for over thirty years.

345

June Alaniz

A SACRED PLACE

Tall Sleeping Giants swaying in the wind
I look up and see a bright green canopy of leaves.
It's 5pm, but I can notice a breeze in this shade. I smile, pleased.
Long gentle curved hands wave hello.
It was a hard day at work today.
But for a moment... I feel relaxed.
I breathe in deeply
Unhesitant to the air I am breathing.
It enters my lungs freely—no coughing or wheezing.
I am grateful for my Giant Friends
And every way they benefit me.

Alex Hiam   *Swallows*

Lauren E. Oakes

FOREST CLEARING TO GREENING, WITH DREAMS FOR FUTURE LIFE

Nature imitates art, and it wouldn't surprise me if someone
actually found an elderly shepherd who'd spent his life reforesting
whole landscapes. In fact, I'm sure he does exist.[1]

— Aline Giono, 1975

A somewhat hidden figure in the recent history of our world's forests, their comings and goings at the hand of man—that's how I think of Alexander Mather. The portrait I've been able to assemble is of a caring and insightful man, a bit dour yet dedicated and unassuming. The northeast corner of Scotland, where "Sandy" was born in 1943 and where he stayed, is not particularly famed for a culture of openness, and those who knew him say he was no exception. But he was driven in his research and remarkably productive. He spent his life working on what some scientists today note as a topic that was once fairly academic and has now become one of the most pressing of our time. He called it the "forest transition"—the intriguing and unexpected shift from shrinking forest area to expanding forest area, from clearing to planting trees.

"The shrinking of the world forest is one of the major environmental issues of the day," he wrote in 1992. "For the first time in history, concern about the forest resource has become

a global—as opposed to a national or regional—issue."[2] The world was focused on the vanishing forests, on the increasing rates of deforestation in the tropics and beyond, but Sandy Mather was looking another direction. He was captivated by the greening—reforestation, the planting of trees on cleared lands, and afforestation, the planting of trees on land that was not previously forest, or expansion through natural regeneration.

By the late 1900s, forest area in Denmark had almost tripled since 1800. In Switzerland, it shifted from 19 percent to 32 percent between the 1860s and 1980s. One of the earliest and most extensive transitions occurred in Japan in the 18th century. In New Zealand, forests and woodland area increased by about 40 percent between the mid 1960s and 1980s.[3] And the Scottish countryside offered another case for Mather to study—this one, closer to home. Mather left his mark in the many articles and books he wrote to unpack how and why such historical forest transitions occur.

Stu Smith *Seaton Park, Aberdeen*

"I think his interest stemmed from his feeling for the landscape and environment of Scotland," one researcher told me, from witnessing "how people had interacted with and influenced it over time." A scientist observing patterns in place instinctively looks outward, searching to find the limits of generalizability. "Paradoxically, the more separated people become from the forest in their daily lives," Mather writes, "the more interested and concerned they seem to become about its condition and extent."[4] Walks in the countryside outside Aberdeen, in the hills and among the trees, were his refuge.

I first came across Mather's work because I, too, wanted to know what had sparked such forest transitions of the past. I wanted (and still want) to know what could spark one today at the rate and scale needed to harness the

potential of trees to help combat climate change.

Intact ecosystems, especially our vast tropical and boreal forests, soak up around 30 percent of mankind's carbon emissions each year. They can continue to provide that much-needed benefit if effectively protected—indeed, that's one of the many arguments against deforestation today. But some scientists today say the earth could hold another 1.2 trillion trees, and planting more forests could absorb another two hundred gigatons of carbon. That's about two decades of global carbon emissions at the current rate.

"Saving the forests has become a major concern in the late twentieth century," Mather noted thirty years ago, "not only for their resource value but also for the environmental benefits they provide."[5] The concerns motivating action back then weren't only centered

on the push for carbon, but today they are increasingly so. Add carbon sequestration to the long list of many values we're striving to sustain: forest-dependent species, water quality and quantity, wood products, and much more from the mosaic of lingering green. There are the intact forests that remain, the degraded ones we can restore, the plantations we cut and replant for wood products, and perhaps the forests of the future, where a new type of forest management may become farming for carbon.

Despite the increasing global interest in reforestation, forest losses are still much greater than gains. The number of countries showing increases in forest cover is relatively small—on the order of sixteen to twenty. But that doesn't mean the reversals aren't happening at smaller scales, as plantings scatter throughout landscapes and communities across the planet.

Seed collected and cultivated. Tender seedlings prized and protected. Hands together, cupping the roots of the trees for tomorrow. Planting may be an act of fighting for an inhabitable planet, a world where forests offer some long-term insurance but only to complement what must also come from cutting emissions themselves. There are citizens and leaders seeking to recreate what was lost and perhaps to renovate—to repair or improve, to remodel, to impart new vigor, to revive.

Mather and his colleagues saw two pathways driving the historical forest transitions. In the one driven primarily by economic development, industrialization and agricultural intensification lead to abandonment of less-productive farm lands. Often on hillsides

349

Alli Maloney *Sun rays — Esselen Land, 2017*

surrounding productive valleys, those forsaken farms—left fallow—could revert to forests, and many did. In the other pathway, what Mather called the "crisis-response" model, scientists and policymakers who perceive immediate threats from deforestation respond with conservation and reforestation measures. In Switzerland, he documented fears of floods and wood shortages behind efforts to expand forests. In Denmark, a growing population combined with increasing agricultural land use and demand for wood triggered such a forest-resource crisis by the beginning of the 19th century. Underlying the complexity of driving factors was "a legacy of unsustainable land use."[6] Ensuing problems became a stimulus for action; they helped chart the new course toward a forest transition. Could they do so again, today? And in the years to come?

Part of what was so appealing about

Mather's explanation was the structural clarity (dare I say, the apparent simplicity) of the root causes. But that also welcomed the other researchers who followed to note exceptions and complexities. There's the scarcity of forest products, for example, leading governments and land managers to establish planting programs. Changes in national forest policies can play another role, triggered by the scarcity but also by other interests—to promote tourism or to "green" a country's image.[7] A more modern version of Mather's economic development route has also been observed, as forest conservation activities increasingly take place on private lands in the global economy. And then there are the mosaics of wooded landscapes at the margins of forests—the notable increases of tree cover from landowners expanding fruit orchards, gardens, or hedgerows, or agroforestry. Even such domestic forests, occurring at a much smaller scale than those that Mather deciphered, can (and do) add up.

But perhaps what we are seeing today and dreaming for tomorrow is something new under the sun. Ambitious targets, adopted around the planet, aim for forest expansion in coming decades: the Bonn Challenge, for example, seeks to restore an area equivalent to 30 percent of the world's remaining primary forests by 2030. Countries have made their commitments, too—north to south, east to west, from Hungary to Nigeria, from Guatemala to New Zealand. These are national or global targets, that require local action and the support of citizens in the places where forests could be again. It's the rapid rate, collective action, and massive coordination needed that make these new transitions, in both theory and practice, remarkably different from the transitions Mather studied. But different is

what may be required to deliver the scale of action needed today.

It is "possible that the transitions may occur in a more rapid and abrupt fashion," Mather wrote. Perhaps his words, echoing decades later, are a premonition of all to come. But the native forest, he also noted, was almost completely eliminated in several developed countries before a transition toward expanding forests took place. The risk is acting too late.

Even back then, he keenly observed such forest transitions were unlikely to re-create the conditions of the past. Not all that's lost could be recovered, nor can it be today. Trees alone will not stop the world from warming. But our relationships to the green, shaped by what we choose to protect and what we work together to create, will also determine what lies ahead. Seedlings in soil, canopies in the sky: these are dreams for future life. ○

Notes

1. A. Giono, "Afterward," in J. Giogo, *The Man Who Planted Trees* (England: Harvill Secker, 1996), 40.
2. A. S. Mather, "The Forest Transition," *Area* 24, no. 4 (December 1992): 367–379.
3. Mather, 368.
4. A. S. Mather, *Global Forest Resources* (Oregon: Timber Press, 1990), 1.
5. A. S. Mather, *Afforestation: Policies, Planning and Progress* (England: Belhaven Press, 1993), 1.
6. A. S. Mather, C.L Needle, J.R. Coull, "From resource crisis to sustainability: the forest transition in Denmark," *The International Journal of Sustainable Development & World Ecology* 5, no. 3 (June 1999): 182–193, 191.
7. E. F. Lambin, P. Meyfroidt, "Land use transitions: Socio-ecological feedback versus socio-economic change," *Land Use Policy* 27, no. 2, (April 2010): 108–118.

↑ *Exhibition - drawings from Index of American Design*
The Federal Art Project's Index of American Design was an attempt
at surveying American decorative, domestic, and folk arts from
the colonial era to the beginning of the twentieth century. From 1936
to 1942, nearly one thousand artists in thirty-four states and the
District of Columbia were employed to document some eighteen
thousand ordinary objects, with the aim of creating an illustrated
record of the history of American design. This poster advertising an
early exhibition of drawings from the Index shows a weather vane
in the shape of a Native American.
→ **Seán Ó Domhnaill**   *A Cattle Farm to the North of Ullapool*

Invitations

Voting, Research Stations, USPS

Jamie Hunyor

LINES AT THE END OF THE SEASON

I went away looking for a farmer's solitude. I went to the yellowed West. I lived in a shed just big enough to hold me. In the fields the sun beat down on vegetable scraps. I was no hero. Poking into the cradle of darkness I clung tight to the hand of another. I felt the frustration of possibilities and blind growth like the probing pain of an infant's first set of teeth. I wrote a few poems, dirtied and cleaned the space where I lived, dirtied and cleaned some more. I dreamed of rain, fat droplets falling into the dusty hands of a gravel road in Athens, Ohio. In a few years' time not much will be left but the lingering smells of pig shit and six or seven overripe peaches deflating on the kitchen table. ○

← **Rose Robinson** *Vermont*

From the *Larned Chronoscope*, published under that
name from 1907–1930. The small town of Larned, Kansas,
was built at the confluence of the Pawnee and the
Arkansas Rivers.

LARNED CHRONOSCOPE

WOLCOTT & CHRISTY.

Harry H. Wolcott.   Lynn M. Christy.

THURSDAY, JUNE 17, 1909.

The Joys of Farm Life.

People talk about the drudgery of farm life. I do not want to paint it with a deceptive rose color. There is plenty of hard work on a farm. But then it is so interesting! Here, just as with any other kind of work, it is immensely important to plan your work intelligently, says a writer in the New Idea Woman's Magazine. When the hard special tasks come along, the spring planting, the haying, and harvesting, and threshing, and butchering, and getting in ice, it is possible to approach them in such a way that the household feels each one almost as a festival. We all work hard at these times, but we enjoy them.

Although we are so busy, William and I try not to be too absorbed to heed the beauty on all sides of us. We think that our little farm has a particularly fair outlook upon rolling fields and woodlands, our little brook, a far rampart of blue mountains, but nature everywhere, in every season, is beautiful and refreshing. We feel that we and our children ought to be better and happier for this constant influence of beauty.

Ours is the really simple life. We are kept close to the great simple facts of life and growth. All the forces of physical science are at work in our little farm. We can succeed only by obedience to the law of nature. Thus one may learn serene docility. On the farm our wants remain simple. They spring from our real needs. They are not created by gazing covetously upon the possessions of others. No occupation can make a better use than farming of trained intelligence, or of a greater variety of information. Everything I ever learned has been of use here. The constant variety of task is what makes farm work so interesting. The seasons and the conditions are constantly changing, and the farmer's tasks with them. When people talk about the dullness and monotony of farm life, I simply cannot understand what they mean.

The farm is the best possible school for developing individuality. It calls continually for all the initiative and ingenuity and resourcefulness that you can possible muster. One of its great rewards is a joyous sense of power. It is a delight to feel yourself developing, and creating, and getting all that you possibly can out of your own little portion of good old Mother Earth.

David Boyle

DISTRIBUTISM, BACK TO THE LAND, AND THE LEADERSHIP OF WOMEN

The great coronavirus pandemic has focused people's minds on health, income, and, in particular, on food. Where I live in the English South Downs, what food has remained consistently available has tended to be local, with short trade chains. I feel sure that sense of reliability will carry on once the pandemic has passed. In fact, I will not be at all surprised if the aftermath of this virus inspires a revival of the English back-to-the-land tradition.

"Back-to-the-land" is one of southeast England's contributions to the world. A reaction against the industrialization of London, the movement came of age in the early 19th century and was shaped by a group of thinkers including William Cobbett (a radical), Samuel Palmer (a conservative), William Morris and Richard Jefferies (socialists), John Ruskin (a high Tory, or so he said), Peter Kropotkin (a communist), Mahatma Gandhi (an Indian nationalist), G. K. Chesterton and Hilaire Belloc (both liberals and then distributists), Henry Williamson (a fascist), Fritz Schumacher (green?), and so on.

These may not have been the political labels they would have chosen to describe themselves, but all of these thinkers influenced each other in a tradition that stretches through to today, when you might include people like Women's Police Volunteers Founder Nina Boyle (a distributist but only a very distant relative) or organics pioneer Eve Balfour. The tradition emerged from the United States: Thomas Jefferson had great influence on Cobbett during his period there as a young man. Through Gandhi, Pope Leo XIII, and Kropotkin, it reached the rest of the world.

Most, though not all, of these thinkers also had links to the high Anglican or Roman Catholic Church (even if just through an obsession

357

with Gothic architecture). Many held conservative views on women that conflicted with the radical elements of their philosophies. This was particularly an issue for the distributists, the followers of Belloc and Chesterton in the 1920s and 1930s, whose anti-corporate radicalism belied an ultra-conservative streak. Both Belloc and Chesterton opposed women's suffrage—and like many of the other back-to-the-land pioneers, they knew how to use a pen rather more than they knew how to use a spade.

They were great wordsmiths. Their defense of small-scale and rural was wonderful but I don't believe they expected to succeed. "Do anything, however small, that will delay the work of capitalist combination," Chesterton wrote in 1926. "Save one shop out of a hundred shops. Save one croft out of a hundred crofts. Keep open one door out of a hundred doors; for so long as one door is open, we are not in prison."

It is wonderful stuff, but it isn't the rhetoric of victory.

Perhaps that's why the distributists failed—they clearly expected to. As I was going to explain in a lecture at the Ditchling Museum in

Dorothy Day, editor of the Catholic Worker
Courtesy of Jim Forest

Sussex—cancelled because of the virus—they were melancholic theorists rather than optimistic doers.

Things have changed over the past generation, which means that back-to-the-land may once again become a political factor, at least where I live. The renewal of this movement has an obvious relevance to the climate crisis, the need for local regeneration, and the means to achieve it. But women are now the driving forces.

The American distributist Dorothy Day, editor of the *Catholic Worker*, marked a shift from predominantly male to predominantly female leadership in the back-to-the-land movement. Day passed away in 1980 and those who've followed—Pam Warhurst of Incredible Edible in Todmorden; Green Belt founder Wangari Maathai in Kenya; the pioneers of the local food movement across New England; the tree-hugging Chipko of Uttar Pradesh—have tended to be women.

It was the early distributists who led the way in battling against eugenics, agricultural animal cruelty, and pesticides. But while

Catholic Worker, Vol. XXVIII No.3, October 1961
Courtesy of Jim Forest

→ **William Morris** *Fruit or Pomegranate*

Ruskin wrote and Palmer painted, the women are acting. Gandhi's program of pairing agrarian devolution with nation-rebuilding, in some ways an offshoot of the same English tradition, is now often led by women (as in Wangari Maathai).

One highly visible indicator of this shift is the change in the status of allotments in the United Kingdom. By the 1970s, cultivating these small urban plots tended to be a dwindling retirement activity for men of a certain age, like pigeon-fancying. By 2010, there were one-hundred thousand people waiting for allotments in London alone. Those who are growing broccoli and yams on what have become extraordinary multi-ethnic landscapes are overwhelmingly women.

The fact that these are women may not be as important as that they are intensely practical people—people who understand the importance of food, the resilience of local supply chains, and the value of digging in the dirt.

I hope, once the virus has moved on, or become normalized, I may be asked back to give the lecture. The Ditchling Museum is a kind

Mark Garten *Wangari Maathai, founder of the Green Belt Movement*

Women of the Big Chipko movement

of ground zero for distributists thanks to the Guild of St Dominic and St Joseph, which, ahead of World War I, shaped a small Sussex town into the center of a political approach to living and the world—one based on arts, crafts, rural life, and religious contemplation. It is something to look forward to. ○

Rob McCall

AWANADJO ALMANACK

An Invitation

Rank opinion—
In years past, we would see Summer people winding down their activities in early fall and getting ready to go back home. In 2020, the word around town was that many of our Summer visitors were going to stay a while— maybe a long while. With jobs and schools being carried out largely online, many were saying, *Why go back?* This could be a blessing to us IF you are ready to pitch in and help do what needs to be done around here: Help our volunteer fire departments do their jobs, help keep our rural hospitals open, help produce local food, help struggling fishermen and others find good work, help our small businesses survive. If you stay, help us strengthen our towns and institutions and help us take care of each other in this ancient and fine way of life.

This was originally published in Awanadjo Almanack *for the week of August 28 – September 4, 2020. The* Awanadjo Almanack *appears in the* Quoddy Tides, The Weekly Packet, *and* Maine Boats, Homes & Harbors, *and as a podcast on* WERU *Community Radio.*

Alison Chow Palm

TRAUMA AND HEALING

Trauma is harm that clings. It separates us, scrambles memories, and disrupts communication—between parts of ourselves, between individuals and communities, between species. How do we turn from destructive patterns of coping towards dreaming new dreams and creating healthy futures?

Emotional trauma during childhood has been shown to increase health problems later in life, such as heart disease, asthma, anxiety, depression, and alcohol and drug misuse. When trauma is caused by years of interaction with abusive individuals or unjust systems, we learn to survive within the framework, normalizing perpetual harm. Trauma can cause individuals to shut down communication between parts of the brain—especially in periods of stress—and can fragment and distort memory. In my work as an emergency nurse, I see the impacts of people's pasts played out daily: the teenager brought in after an overdose whose parents arrive and berate them, the man with a new diabetes diagnosis whose finances will compel him to continue eating spaghetti sandwiches.

Cultural trauma and structural violence are wounds in the stories of who we are, cutting us off from what we know, what we love, and where we come from. These wounds shape our dreams and interactions, creating a baseline of abuse. These kinds of wounds erase our collective memory of what it is to be human, and what it means to have a relationship to place.

This is the kind of harm used to enforce imperialism and enslavement. For my family,

the government-created Cultural Revolution in China created silence. Silence that hid our family's Buddhism from my father's generation, silence when my grandparents disappeared to faraway engineering jobs, silence that built on thousands of years of oppression when my great-grandmother with bound feet couldn't walk fast enough to catch up with the children. Silence that left a family without the support of the knowledge and caring of our ancestors.

Ecosystem and landscape trauma is also caused by violence and force. When living beings are exterminated, interspecies relationships disappear. When members of long-lived species are killed, or prevented from growing, and their communication pathways are removed, community memory and communal sharing of nutrients are destroyed. Harm can be caused by synthetic inputs that change chemical proportions and that kill plants,

insects, and fungi. Landscape change caused by humans can influence how water, sunlight, and wind interact with living beings. Denying that humans live in a many-species community is part of the separation that leads to dissonance. A place I love spent years in monoculture cultivation enforced by pesticide use and heavy tillage. It's been a decade since that last dusty year, and the healing is palpable. Yet, compared to the presence and complexity of old-growth forest, the damage and unbalance is enormous.

Opportunities for healing can be created by removing harm and upholding safety, for ecosystems, individuals, and communities. This means acknowledging and ending sources of pain, decolonizing our lives, denormalizing violence, and supporting Indigenous work to protect land and culture. This could also be leaving a relationship, removing bans on who may marry whom, or bringing an end to clear-cut logging.

Processing is crucial. This could be grieving, doing somatic therapy, talking with our families and communities about our narratives, giving fallow time, or allowing secession on a damaged piece of land.

We can encourage space for creation and maintenance of relationships, communication, and unbroken memory. For our brains, that means learning to find new neural pathways. For our human relationships, that means being mindful of what we say and how we interact. For the land, that means creating spaces where we encourage species with long lifespans to live, managing water flow in ecologically

appropriate ways, learning how various species communicate, using cultivation techniques that encourage healthy mycorrhizal networks, and being mindful of where and how we utilize disruption.

Dream beautiful dreams, live healthy patterns. What would our lives and interactions look like if we weren't afraid of systems of harm? What would it feel like to be part of an intact ecosystem? Who can we learn from? What stories and knowledge help us live the way we want to? What world do we want to wake up in every morning? ○

Addison Bowe

The Submergence Collective

SCORES FOR RESTORATION

Watch it go
Chart the flow of water
on the land.
Squint; where has it moved
before?
Where will it go next?

Offer a seed
to its wending path.
Watch it go.

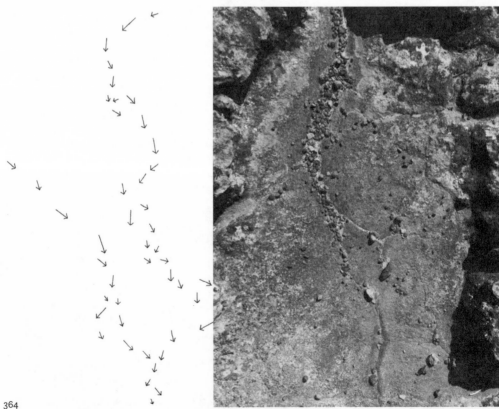

Someone you feel far away from
Lay on a slab of concrete. Whisper

to the soil underneath
about someone you feel far away from

Lay there until your sweat saturates the concrete.

Imagine the smell of a summer rain on hot concrete
pouring onto the city
and evaporating

Then water a crack in the concrete
until it infiltrates the soil.

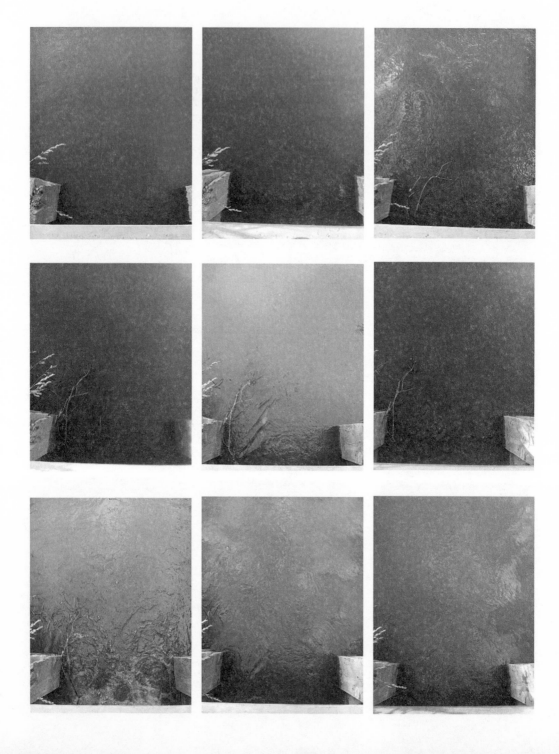

How does the light fall on their body?
Go to the nearest body of water
introduce yourself to them

return at the same time (everyday)
form a relationship with them

where do they come from?
how does light fall on their body?
how does this information connect you two?

give a gift of nutrients for the water to carry

Let it lean on you
Where is the wind blowing from today?
Walk into it until you meet a tree
Sit with your back against the tree and let it lean on you

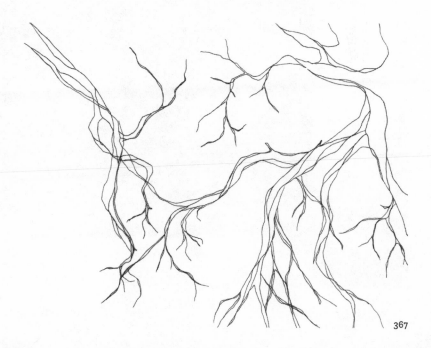

Becoming soil

A) When you wake up, imagine you are crawling from a bed of soil. How would your life be different if you slept in soil?

B) Go outside and try to find the best soil you can. Make note of why you think the soil you chose is the best. Put a good amount in a bucket to carry with you for the day.

C) Spread some soil on your neck, and armpits, for perfume.

D) Dump some soil in your bathtub.

E) Wash your face with some soil.

F) Season your dinner with some soil.

G) Give some soil to your lover or friend instead of a kiss.

H) Give some soil to someone who really needs it.

I) Take a nap in some soil.

J) "Soil" your clothes.

K) Really look at your fingernails.

L) If you have any soil left, take it back to its home and thank it thoroughly. Consider visiting it again tomorrow.

A place of trauma
Go to a place of trauma
Introduce yourself.

walk within. Tell this place a story
speak

o u t l o u d

then, listen.
what does this place say back to you?

offer it your breath—put it in the ground.

Birth

Spend a sunrise or a moonrise watching what you can see
make or unmake themselves. Take the bits you love
and write them on the palms of your hand in lists.
The next day or night, create something or help something create
itself
Let it sit in the sun or moon for a bit
Touch it fondly and smear earth on it
Tell it:
"you belong"
"you belong"
"you belong"
Wash your hands until your lists dissolve, then pat your hands on
the ground with rhythm and
celebration around the thing and say:

"I can't wait to see what you become. I am so happy you are
here. I am so excited to exist on this planet at the same time as
you."

Shamu Sadeh, Janna Siller, and Rebecca Bloomfield

BARREL INSPECTIONS AND FAKE PROFESSIONAL SOCIETIES

How to Entertain Oneself While Remaking the Food System

Eleven young people dismount from the old bicycles that were issued to them yesterday. Pedals scrape ankles and front wheels bump shins as they jostle to heap the bikes against the backside of a gray, prefabricated building that has just been introduced to them as "The Burbs." Surrounded by forested slopes that don't feel a bit like the suburbs, the group has yet to see the actual farm they signed up to work on for three months. They turn expectantly to Shamu Sadeh, founder of the Adamah fellowship program whose website promises a "transformative experience" and site tour leader who optimistically assumed that eleven people could dismount rickety bicycles at the same time in a fifteen-square-foot driveway. He assures them that they'll make it out to the farm soon, but first, he says excitedly, "Welcome to the Center for Cultural Proliferation, where we preserve food and ancestral traditions through symbiotic relationships with bacteria: May we all proliferate. 'Be fruitful and multiply'—as it is written!"

The new fellows enter the commercial kitchen, which looks sterile: shiny steel work surfaces, a walk-in cooler, tall blue pickle barrels. They stand awkwardly, wiggling their toes, wondering if the farm boots they borrowed from a cousin were the right choice.

They peek sideways at the person next to them, wondering how they'll get along living and working with someone so much more, or so much less, Jewish (or spiritual, or "outdoorsy," or Queer, or "normal") than they are, when a person wearing swim goggles, a headlamp, and an apron pops out of one of the pickle barrels holding a magnifying glass and a clipboard.

"Oh hello! So sorry, I wasn't expecting you!" the barrel inhabitant says.

"Oh hello, Cole!" Sadeh says nonchalantly. "I didn't realize it was barrel inspection day."

How can we use jokes to fuel the hyperlocal land work that amounts to a Green New Deal? How do we help young people who are alienated from the sources that sustain them get comfortable with holding shovels and the holy, unrelenting work of schlepping stuff from one place to another? How do those of us in it for the long haul entertain ourselves after we've spent fifteen years weed whacking around the same blueberry bushes, reminding a rotating mix of people not to step in the beds, pulling what feels like the same plastic bags and aluminum foil out of the compost piles, repairing the same leaky irrigation risers (okay, maybe it is a different one every year)?

On our farm we find that humor and, more broadly, imagination, are vehicles that create the world that can/would/should be. We stage mock pickle-barrel inspections. Our to-do list includes a note to call a meeting of the Sauerkraut Advancement Board rather than "discuss marketing." Like our compost piles and our "flerd" (a combination of a flock and a herd) of sheep and goats that eat roadside vines along the edges of our veggie fields, jokes are tools of transformation and regeneration.

Adamah means earth, with roots in the Hebrew name of the first human, Adam. The farm and the fellowship program were created at the Isabella Freedman Jewish Retreat Center in Falls Village, Connecticut, to integrate intentional community, food production, new-old Jewish spirituality, and ecological literacy.

Adamah grows organic food for the local community and retreat center guests, but it is also a place of experimentation. Think chestnut hybrids and food forests, intercropping and reduced tillage. Adamah is a place for the Jewish community to connect to its roots, express gratitude, experience connectivity and interdependence, examine what it means to live in diaspora on stolen land, and to reform agricultural systems that are rooted in stolen labor and continue to rely on exploitation.

We, the writers of this treatise on farm humor, are the full-time farm and programming staff, and along with several fellowship alumni who join us as paid apprentices each season, we welcome individuals from a wide variety of backgrounds to coax food from our loamy Berkshire slopes. Some Adamah fellows—or, as we call them "Adamahniks"—have received their entire education from Brooklyn *yeshivas,* or religious Jewish educational institutions. Others have grown up Episcopalian in Ohio, and recently discovered a Jewish ancestor whose tradition they want to connect to. Folks come here from all kinds

of Jewish backgrounds, finding connection through traditions whether they evolved in Russia, Syria, Hungary, Morocco, or elsewhere. Adamahniks may have spent several seasons interning on production farms, or they may only have stepped off of concrete a handful of times in their lives. Judaism may be a law-based religion with a firm canon to some while, to others, it may be a sensual, shape-shifting experience of malleable ritual.

One thing everyone has in common when they arrive at Adamah is that they are, to some degree, uncomfortable and disoriented. We help them feel welcome with smiles, hugs, and assurances that they are in the right place. Fresh farm veggies and last year's sauerkraut help, as does an extremely detailed schedule that makes clear where they are supposed to be at what time and what they are supposed to do once there. We simultaneously dive straight into the disorientation by using confusing names for things and places. It is surreal enough that adults have to spend months outside their "normal" lives to learn how to produce the food that sustains them every day—why gloss over the dreamlikeness of reality by calling a computer a computer and the fourth of July, "Independence Day?" Here, it's a "confuser" and "Interdependence Day."

In the religious experience and in experiential education theory, disorientation is the precursor to a transformative shift in how we perceive the world. At Adamah, we use cultural disorientation to help participants "lift their eyes" and see the new, and very old, wells of potential that are dug deep before us.

American society is so disconnected from community land care that our language has no word for the place on a farm where you keep dozens of buckets hanging individually on wooden posts (please, g-ddess forbid, don't

Alejo Salcedo *Bull and Boy*

stack them—they'll end up stuck together for eternity!). We call it Bucketon Acres. In the fall, when leaf blowers across the country whirr in an effort to stuff garbage bags full of the carbon that trees spent the summer manufacturing out of thin air, Adamahniks participate in the "No Leaf Left Behind" campaign to schlep tarps full of leaves from the lawns of the town library, the retired kindergarten teacher's home, and the retreat center campus to our compost piles. We open boring meetings about scheduling and logistics with a game show called "What's in this Bucket?!" The winners score prizes like the uncorked bottle of wine found mysteriously unbroken in the compost.

We routinely stage pickle barrel pop-outs because if we try to address inequity in our food system, face anti-Semitism and other forms of persecution, and tackle the climate crisis sans humor, we'll be crushed under the weight. Humor has to work delicately because it can be as powerful a trigger of trauma as it can be a medicine. For us, jokes are worth the time and care because people are more likely to join us in clearing poison ivy off of the deer fence if we tell them they are part of the Professional Society for Mammalian Herbivore Exclusion and Rash Prevention (PSMHERP).

So, dear farmers, this season while you are building the future by setting seed to soil, and in between calling your governor on repeat to encourage them to embrace regenerative farming in your state's climate action plan, we hope you'll also take a minute to prank call your fellow farmers. Silly voices. Voicemails confirming the delivery they placed for macro-biotically treated quail manure. Fake apprenticeship applications from budding young farmers who promise to increase your yields with homemade tinctures of apple cider vinegar, hamster urine, CBD, and hemlock. Stash some costumes in the barn and display drip tape figurines in the office. Just remember that you get to laugh while you re-work the global food economy with your bare hands. ○

Debbie Hillman

URBAN SOILS,
URBAN CONSCIOUSNESS

An urban habitat will never grow all of its own food. It will always have more people, buildings, and hardscape than farmland or arable soil. Even Havana, Cuba, a model of urban agriculture for forty years, grows only about 50 percent of its food. But people living in cities and suburbs—currently 70 percent of the US population—can achieve 100 percent soil consciousness.

Why soil? Because soil is the medium that grows the earth's green skin—plants— the primary source of our food, oxygen, and freshwater. The best way to achieve soil consciousness is to think like an organic farmer. Or, to use the latest terminology, like an agroecologist.

Thinking Like an Agroecologist

In May 2019, Dr. Jonathan Lundgren, an agroecologist based in Estelline, South Dakota, gave a talk about his work to restore the vitality of agriculture production systems by modeling nature. He presented the following principles along with a call to action for farmers.

Principles to farm by
—Stop tilling (or reduce it)
—Never leave bare soil
—Some plant diversity is better than none, more is better than less
—Integrate crops and livestock

Call to action
The decisions you make on your farms transcend the boundaries of your farms. The question you should ask yourself every day is, Why am I doing what I'm doing?[1]

Call to Action for US Cities & Suburbs

These principles can be applied in urban and suburban areas by property owners, homeowners, lawn service companies, landscape designers, business associations, religious and educational institutions, voters, officials— anyone concerned about biodiversity, topsoil, food sovereignty, and meaningful living.

The decisions you and your neighbors make on your own property, the land in your neighborhood, and all the land of your town transcend their boundaries. As you walk over the land of your urban or suburban community, ask yourself or your neighbors, *Why do we throw out our leaves every fall? Why do we rake our beds bare in the spring? Why do we bag up our grass clippings? Why do we chase every leaf particle and twig down the sidewalk with noisy, gasoline-powered leafblowers? Why do we spend time and money to destroy our soils?*

Or, on the rare occasion you come across the opposite behavior, you might ask yourself or your neighbor, *Why do you leave your leaves, your grass clippings, your dried flower stems, seed stalks, grasses, and other green matter?*

Pierre Joseph Hubert Cuypers *Impressions of Leaves*

After twenty-five years as a professional gardener and forty-three years in one very urban suburb, I have found only one answer to the first question: Neatness.

There are a myriad of answers to the second. Leaving the leaves will ...

—Create new soil, adding structure and nutrients, improving growing conditions— for free.

—Protect existing soil from wind erosion.

—Hold moisture in the soil; plants will need less water, less frequently.

—Protect plants against extreme cold or heat.

—Give you and your neighbors peace and quiet by decreased use of leaf blowers.

—Keep our air cleaner with less fumes, less stirred-up dust.

—Decrease fossil fuel use with less frequent use of leafblowers, gasoline-powered lawn mowers, and trucking of materials back and forth.

—Eliminate much of the time and cost of bagging up leaves.

—Protect the habitat of the small creatures— worms, insects, microorganisms.

—Show us the beauty of natural processes.

—Produce abundant, nutritious food.

For the last forty-odd years, my hometown (like most other US cities and suburbs) has thrown out 90 percent of our land's production, regularly ripping apart the web of life—week by week, year by year. It is no accident that during that time our human social fabric has also been shredded.

Time for us all to start thinking like agroecologists. ○

Note

1. Thanks to Mallory Krieger of The Land Connection for sharing her notes from the University of Illinois at Urbana-Champaign event over the Regeneration Midwest listserv.

Dan Kelly

ON PATHS MOVING FORWARD

Food Sovereignty in a Colonized State:
A Dialogue with Māori Organics Practitioner Dr. Jessica Hutchings

As the global movement for food sovereignty continues to gain momentum, it's become increasingly important for those of us in Aotearoa New Zealand to acknowledge both our overlaps, and our points of difference. In these matters, there are times when it is best not to speak, but to listen. For me, writing as a *Pākehā*—a descendent of European settlers—such listening is to open up the conversation, to move beyond the modern ills of industrial agriculture and into the older ones that birthed them, those same processes of division and control that now farm fields of corn, once used to clear people and claim their land.

What follows is an interview with Dr. Jessica Hutchings, a *Hua Parakore* (Māori organics) practitioner and author of *Te Mahi Māra Hua Parakore: A Māori Food Sovereignty Handbook*, exploring these issues as they relate to the lands that we both call home. However, before we jump in, I want to give a brief primer on the situation in New Zealand, and the history that informs our paths moving forward.

While overseas versions of food sovereignty have focused on peasant rights and their ability to choose what food is grown and where, in the context of a colonized country like Aotearoa New Zealand, the use of "sovereignty" has a different, more painful history. The erosion of

tino rangatiratanga (self-determination, sovereignty) exercised by Māori, the Indigenous peoples of Aotearoa New Zealand, prior to colonization is both a breach of international human rights law, and the more local agreement now widely considered to be the founding document of New Zealand: *Te Tiriti o Waitangi*, or the Treaty of Waitangi. Signed in 1840, Te Tiriti is a compact between Māori and the British Crown in which the Crown recognized the existing sovereignty of Māori, promising Māori the rights of British subjects and the continued chieftainship over their lands, villages, and treasures in exchange for the exclusive right to purchase land. It is, put simply, the agreement to which settler descendants like myself trace our rights to belong in this land—not without accountability, to the detriment of those here before, but in relationship, working to uphold and restore that which was promised in the moment we claim as starting "us" off. As Māori academic and lawyer Moana Jackson often explains, treaties aren't something you settle, but honor. For Pākehā, this means facing up to our past, and all our uncomfortable history has done to undermine Māori-Pākehā relationships. As our gardens reveal, nothing is ever the same. Things change and grow and the possibility for honor remains.

For food sovereignty to thrive in both New Zealand and other colonized lands, practitioners will need to face this task, treading carefully and with open ears, less the exclusions before be remade.

Kia ora Jessica, I want to start off by thanking you for your time today, and all the work that's led up to this—for those who aren't familiar, can you give us some background about who you are and what you do?
Tēnā koe, I am from Ngāi Tahu and from Gujarat, India. I am passionate about *Papatūā-nuku* (the Earth Mother) and the self-determining rights of Indigenous peoples to live in ways that are connected to both land and cosmos. I live on a small *whānau* (family) farm in Kaitoke, just north of Upper Hutt, where we grow *kai* (food) to feed whānau, operate within a closed loop system as much as possible and work with both Hua Parakore (Māori organic) practices alongside biodynamics. I have been growing and eating kai from our *whenua* (land) for the last fifteen years.

My personal and my professional lives are connected through *kaupapa* (issues) that I commit to. I am trained as a kaupapa Māori researcher and have worked in the Māori environment, science, and development sectors for the last two decades. I am driven by contributing to the development of flourishing Māori food communities which, of course, is about Māori food sovereignty and, for me, about Māori organics.

To someone who hasn't heard about the food sovereignty movement, how would you explain the need for a shift? What is it about our modern food system that is so problematic?
Simply put, our current food system is broken. It is predicated on a global capitalist model that circulates around profit, productivity, and efficiency, not around providing equitable access to all to safe, nutritious, and culturally appropriate food. This global food system responds to drivers that are not set by consumers but by those who seek to profit. It drives monocultures in agriculture which then show up in our food system in highly processed foods that are based on wheat, rice, and soya—much of which the seed companies own the rights to.

Food sovereignty seems to be one of those terms that has a range of meanings, with different practitioners focusing on different aspects. Some readers may be familiar with La Via Campesina, the International Peasants' Movement. In 2007, a gathering of five hundred La Via Campesina representatives from eighty countries issued the Nyéléni Declaration, organized around six guiding principles. For them, food sovereignty focuses on food for people, values food providers, localizes food systems, puts control locally, builds knowledge and skills, and works with nature. However, while such a definition is useful for building an inclusive movement, it has also been critiqued for being too broad. How would you define food sovereignty?
Food sovereignty for me has to take into account our standpoint. We need to first ask the simple questions: Whose land are we on? And on whose land does our food grow? We often do not see the cultural landscape of Aotearoa due to its erosion through settler-colonial processes. Māori food sovereignty pulls food sovereignty out of the mainstream or white, middle-class discourse and into an Indigenous and kaupapa Māori paradigm. At the heart of it, it is about creating food-secure futures for whānau where we can access safe, nutritious food; it is also about returning to eat

Pedro Szekely *New Zealand countryside*

the landscapes from which we come from, and through this, being able to continue to build our Māori food stories and Indigenous narratives to place and space.

I think that question—whose land are we on?—is crucial. The food has to come from somewhere! While "sovereignty," or *rangatiratanga*, can be interpreted in a number of ways, it's important to acknowledge our past, the Māori sovereignty recognized in Ti Tiriti o Waitangi, and the ways in which this has been undermined by the state. Despite widely held misperceptions, this sovereignty has never been ceded. While Māori have suffered at the hands of a number of incredibly duplicitous and racist policies throughout the years, there has also been staunch resistance and solidarity. Do you see food sovereignty as a way to continue this resistance? What support would you like to see from Pākehā and *tauiwi* in this space? Of course, this is what frustrates me about white discourses of food sovereignty. There is often a persisting, settler-colonial erasure of Māori connection to place and the theft of our Indigenous lands and food systems—our ability to eat the landscapes of our *tupuna* (ancestors) and our relationship with food has been dramatically fractured because of colonization.

Having Pākehā who can be allies in this space is important. Pākehā who can talk to colonial injustice and know the history of this land is critical in moving to a healing space. With the rapid rise of provenance in food stories, we need to ensure that Māori are included here and that the New Zealand food story does not perpetuate the settler-colonial erasure of Māori. We need to (re-)story *kaitiakitanga* (guardianship) into our landscapes, agricultural practices, and food stories. For this to happen, Māori

need to be the ones telling the stories. Food provenance in Aotearoa New Zealand needs to not perpetuate colonization but be part of the healing processes inherent in decolonization.

Māori farmer plowing for turnip sowing, 1947

As you explain in your book, food sovereignty isn't a novel concept for Indigenous peoples. Prior to the arrival of Europeans, and for many years after, Māori grew, gathered, and hunted all the food they needed, not just for survival, but for feasts, trade, and hosting. Could you tell us a little bit about these histories, and the worldview that informs them?
Our histories are embedded in land and waterscapes in both Aotearoa and *te Moana Nui A Kiwa* (the Pacific Ocean). Everything is interconnected and based on *whakapapa* (genealogy), from earth to cosmos where *tangata* (people) are not seen as separate from nature but a part of nature, as nature itself. This is very different from other paradigms of thinking, especially that of Western knowledge with its origins deep in the Enlightenment period—premised on the separation of "man" from nature so "man" could take dominion over nature—"tame

Gerrit Willem Dijsselhof *Bare tree on a hillside property*

her like a wild beast" as was said by some Enlightenment thinkers.

In terms of food production, I understand that you practice permaculture, and its overall focus on working with natural systems, rather than in denial of them. For many in the West, permaculture can seem revolutionary—a reflection of just how far we've come from our more involved peasant pasts—but for Indigenous growers, there is nothing particularly novel here: observing and working with natural systems, practicing water retention and improving soil health, co-planting and other 'permaculture' techniques that are ancient and ongoing. Do you see a need to promote this past? How important are these holistic processes—the interconnectedness of natural and human elements—to you, and how do they shape your work?

I am nature, I feel myself as nature. Living in the modern world it can at times be very hard to feel this. What does feeling like nature feel like—how often do we talk about this with others, when do we sense this? I feel very lucky that we live on a small, whānau, food-producing farm, where across the road are ancient indigenous forests and tree beings that transcend time. These daily earthly connections, along with a dedicated practice of yoga in my life, are all manifestations of holistic processes, they reflect the interconnectedness of tangata and whenua.

This interconnection is one that the West has long denied, the perceived division (and assumed superiority that goes with it) giving rise to incredible violence, both against the natural world and Indigenous peoples. While food sovereignty appears, on a superficial level at least, to be concerned with food supply and

381

access, its calls for agency ultimately position it as a challenge to the deep inequalities of power still so pervasive today. As the Nyéléni Declaration makes clear: "Food sovereignty implies new social relations free of oppression and inequality between men and women, peoples, racial groups, social classes, and generations."

Addressing the structural imbalances in Aotearoa will require more than a few gardens. Yet, in the practice of gardening and the diversity there celebrated, in the physicality and shared work, and in the presence that the earth requires of us as part of it (and the humility it fosters), one senses the possibility for healing on multiple levels. Is this something that you've noticed in your own gardening? How would you describe the potential for personal transformation in a space of food sovereignty?

Te Mahi Mara Hua Parakore (Māori organics) is a transformative process, working with the whenua and the soil is deeply healing and connective. For me, it takes me outside everyday microaggressions of colonization to feel and be nature. Māori food sovereignty is more than a few community or *marae* (meeting house) gardens; it is also beyond the glossy magazines that promote middle-class, white food stories full of ethnic diversity and culture. It is about decolonization, it is about reducing structural inequities and racism so that Māori communities and whānau have access to safe, nutritious and culturally appropriate food, and can again return to eat the landscapes of which we come from.

Tēnā koe, I want to acknowledge my positionality; as a writer and researcher, I'm particularly interested in diversity and the varied ways food binds people and place. But, for me, I think the challenge is finding a way to celebrate these newer stories of diversity through food while connecting them to a broader struggle, acknowledging that what we eat is political, that we live in a colonized land, and as you said earlier, to have Māori telling their stories, building a shared understanding and solidarity so that those of us who came after can work with and alongside Māori, helping to establish the structural changes hinted at in the concept of food sovereignty, and ultimately, decolonizing. To this end, is there anything else you'd like to add? What is your vision for Aotearoa in 2040?

I also share the vision of constitutional change in Aotearoa to one of a treaty-based constitution.[2] This is part of the larger framework for Maori food sovereignty and being an activist for change on all levels. I also hold a vision of a genetic-engineering-free Aotearoa where Hua Parakore and organic agriculture can recloak our altered landscapes in greater diversity and connectedness with nature, where we again see ourselves as nature beings.

Ngā mihi nui ki a koe Jessica, all the best for the growing season—may it be a bountiful one! ○

Notes

1. A version of this interview was initially published in *Stone Soup*, a free, independent food-focused magazine. You can read old editions online at stonesoupsyndicate.com.
2. For further reading, see the Independent Working Group on Constitutional Transformation's Report of Matike Mai Aotearoa, January 25, 2016.

"...it is not our farmers, who are in the greatest danger of this species of extravagance; for we look to that class of people as the strongest hold of republican simplicity, industry, and virtue. It is from adventurers, swindlers, broken down traders, —all that rapidly increasing class of idlers, too genteel to work, and too proud to beg, —that we have reason to dread examples of extravagance."

From Mrs. Child, "Hints to Persons of Moderate Fortune," in The American Frugal Housewife, 12th ed. *(Boston, Massachusetts: Carter, Hendee, and Co, 1833), 102.*

Alice Blackwood

NATURE POEM

This poem is a nature poem,
it is deep earth
ink and processed wood;
it knows about being a body,
about spines, spleens
and spiders

This poem is not going for a swim,
it has career aspirations
of waterfall, of being read out loud
in bed, to somebody's anti-poetry partner

This poem is sedimentary,
is shifting sand
from one shore   to another,
is stretching muscles to cloudless blue
and honeyeater branchlets

This poem is pendulous, sagging,
needs to strengthen its core,
write down what it is
grateful for

This poem doesn't know
about the bushfires;
if it did, it wouldn't know what to say
about the deafening heat,
the alchemy,
the ashes

This poem is having a crisis of confidence;
it is a pinball machine,
an invisibility cloak,
a pygmy possum,
a lump of lead

This poem	is yet	to meet	a poem
that is	not		a nature poem.

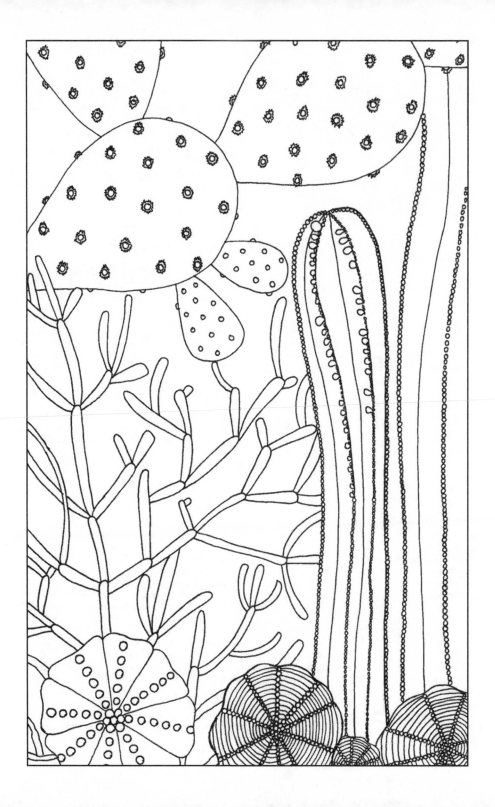

IMAGE CREDITS

This is a noncommmercial, nonprofit publication. To re-print original work, request permission of the contributor.

3 Gerrit Willem Dijsselhof, *Kale boom* (1876 – 1924), Rijksmuseum.

4 Adapted from Richard Nicolaüs Roland Holst, *Ex libris van Adriaan Roland Holst* (1915–1916), Rijksmuseum.

9 Pierre François Legrand after Gerard van Spaendonck, Rose-tremiere, *Alcea rosea* (1799–1801), Rijksmuseum.

11 Original work by Alli Maloney.

12 Original work by Poppy Litchfield.

15 *Demonstration of bomb harpoon killing North Atlantic right whale*, Voyage To Inner Space - Exploring the Seas 1877, NOAA.

16 *Vegetable gardens surrounding the Indian Pueblo of Zuni*, New Mexico, 1873, NARA.

18 Matt Biddulph, *No Dumping, Drains to Ocean*, 2011, Flickr.

19 *Men Drilling for the Passamaquoddy Tidal Power Project*, 1936, Records of the Office of the Chief of Engineers, NARA.

20 Paul VanDerWerf, *Wild Blueberry Bush*, 2015, Flickr.

24 George A. Grant, 1934, NPS.

25 Kai 'Oswald' Seidler, *Rice field in Bali*, 2008, Flickr.

26 *Sketch A Shewing the progress of the Survey in Section No. 1*, 1859, USCS, Geographicus.

28 Original photo by Jon Levitt.

29 Johannes van Loon, *Kaart van de aarde met de verschillende standen van de maan en de zon* (1708), Rijksmuseum.

30,31 Original illustrations by Tatiana Gómez Gaggero.

33 Illustration from *New Ideal Atlas*, 1909, an antique celestial astronomical chart of the phases of the moon, theory of seasons and the solar system. Rawpixel.

34 Donald E Davis, Don E Wilhelms, and US Geological Survey, *Maps of the surface of the Moon*, Library of Congress.

35 Original work by Andrew Stuart.

36 Meagan Racey, *Veazie Dam Removal* (2013), USFWS.

40 *Penobscot River* by Jenni Monet.

43 *Lincoln Pulp and Paper Mill (formerly)* by Jenni Monet.

44 *Maulian Dana, Penobscot Nation's first appointed Ambassador* by Jenni Monet.

47 *Penobscot River* by Jenni Monet.

49 Sherman F. Denton, *The Atlantic Salmon*, Game Birds and Fishes of North America (1856–1937), Rawpixel.

54 Adapted from *No. 7 and 8 steel clipper plow. With jointer, light draft, light weight and strong* (ca. 1870–1900), 19th Century American Trade Cards, Boston Public Library.

55, 56, Adapted from *The 'Planet Jr.' combined drill, wheel hoe, cultivator, rake and plow. Buy 5 tools in one.* (ca. 1870–1900), 19th Century American Trade Cards, Boston Public Library.

58 Original work by Jo Zimny, Flickr.

63 *Alexis Hubert Jaillot, Nova orbis tabula, ad usum serenissimi Burgundiae Ducis,* (1632?-1712), Harvard Map Collection, Harvard University.

64 Master of the Die and Antonio Salamanca, *Plate 22: Venus ordering Psyche to sort a heap of grain, from the 'Fable of Psyche'* (1530–60), Drawings and Prints, Metropolitan Museum of Art.

64 Adapted from Georg Bocskay and Joris Hoefnagel, *Broad Bean and Liverleaf from Mira Calligraphiae Monumenta or The Model Book of Calligraphy* (1561–1596), The Getty Museum.

67 Marion Post Wolcott, *Picking beans near Homestead, Florida* (1939), Farm Security Administration – Office of War Information Photograph Collection, Library of Congress.

71 W.E.B. Du Bois, *A series of statistical charts illustrating the condition of the descendants of former African slaves now in residence in the United States of America* (ca. 1900), Prints and Photographs Division, Library of Congress.

74 *Charleston, West Virginia, Nestled in a river valley in the Allegheny Mountains, the city first attracted weary travelers, who then stayed for the salt, trees, and coal.* (April 3 2020), Astronaut Photography, NASA.

76, 77 J. W. Canfield, *Map of Mahoning County, Ohio: showing the original lots and farm* (1860), Lionel Pincus and Princess Firyal Map Division, New York Public Library.

78 Adapted from J. J. Römer, *Genera insectorum Linnaei et Fabricii iconibus illustrata* (1789), Biodiversity Heritage Library, Smithsonian Libraries.

84 Adapted from *Plate 7 from Carnegie Institution of Washington publication* (1902), Biodiversity Heritage Library, Smithsonian Libraries.

85 Adapted from *Vintage microscopic virus and bacteria illustration*, marinemynt, Rawpixel.

85 Adapted from James Ellworth de Kay, *80.The Hermit Thrush (Merula solitaria)* and *81. The Northern Butcher-bird (Lanius septentrionalis illustration from Zoology of New York* (1842 - 1844), The New York Public Library.

86 Adapted from P.J. Redouté, *Red currant. Ribes rubrum. Choix des plus belles fleurs: et des plus beaux fruits, t. 3,* (1833), Swallowtail Garden Seeds, Flickr.

86 Adapted from *Lauren Bassing Phacelia formulosa USA*, Royal Botanic Garden Sydney.

87 France, Commission des sciences et arts d'Egypte, *Plate 13, Description de l'Égypte* (1809-28), Biodiversity Heritage Library, Smithsonian Libraries.

89 Carol M Highsmith, *What looks like a series of puddles is all that's left, in many spots, of what looks like huge Goose Lake, stretching for miles from northeastern Califor-*

nia into southern Oregon (2012), Prints and Photographs Division, Library of Congress.

98 Albrecht Durer, *The Expulsion from Eden* (1510), National Gallery of Victoria, Melbourne.

102 *Indians farming on Fort Peck Reservation* (between ca. 1915 and ca. 1920), George Grantham Bain Collection, Library of Congress.

105 *Codfish, from Fish from American Waters series (N39) for Allen & Ginter Cigarettes* (1889), The Jefferson R. Burdick Collection, Metropolitan Museum of Art.

106 *Eel, from Fish from American Waters series (N39) for Allen & Ginter Cigarettes* (1889), The Jefferson R. Burdick Collection, Metropolitan Museum of Art.

106 *Tomcod, from Fish from American Waters series (N39) for Allen & Ginter Cigarettes* (1889), The Jefferson R. Burdick Collection, Metropolitan Museum of Art.

110 Aristide de Biseau, *Besneeuwd bospad met jager* (1854–1883), Rijksmuseum.

111 *Sardine Industry, Shore herring weir near Eastport, Me.; the common form of brush weir* (1887), Freshwater and Marine Image Bank, Library of the University of Washington.

112 Lia McLaughlin, *Potter Hill Dam Fishway*, US Fish and Wildlife Service.

115 Emily Vogler, *Set of cards designed for the workshops.*

118, 119 Adapted from Anna Atkins British, *Polysiphonia fruticulosa* (ca. 1853), Gilman Collection, Metropolitan Museum of Art.

120 Gerrit Willem Dijsselhof, *Sketch of fish,* Rijksmuseum.

121 *Toxic Algae Bloom in Lake Erie, Image of the Day for October 5, 2011*, NASA Earth Observatory.

122 *Toxic Algae Bloom in Lake Erie, Image of the Day for September 27, 2017*, Landsat Image Gallery, NASA Earth Observatory.

124 Adapted from *Agnes Chamberlin Trientalis americana* from Original watercolour for "Canadian wildflowers," Agnes Chamberlin papers, Thomas Fisher Rare Book Library, University of Toronto.

133 Illustrations by Lydia Lapporte.

137 Robert Rauschenberg, *Earth Day, April 22 Poster* (1970), Poster for the benefit of the American Environment Foundation, from an edition of 10,300 of which 300 were signed. Published by Castelli Graphics, New York © Robert Rauschenberg Foundation.

139, 140, 141, 142, 143 *On Tidal Zones,* photos by Nick Middleton.

146 *Oyster restoration in Boston Harbor* by A. Frankić.

148, 149 *Wellfleet low tide oyster restoration site* by A. Frankić.

149 Oliver Goldsmith, *Mollusca from A history of the earth and animated nature* (1820), Rawpixel.

150 Adapted from Anna Atkins, *Dictyota atomaria* (ca. 1853), Photographs, Metropolitan Museum of Art.

151 Dorothea Lange, *Pea Pickers Line Up on Edge of Field at Weigh Scale, near Calipatria, Imperial Valley, California,*

February (1939), Photographs, Metropolitan Museum of Art.

153 Jean-François Millet, *The Reaper* (1853), Prints, Metropolitan Museum of Art.

159 Frances Benjamin Johnston, *George Washington Carver, full-length portrait, standing in field, probably at Tuskegee, holding piece of soil* (1906), Prints and Photographs Division, Library of Congress.

160 Archival: George Washington Carver, *How to Grow the Peanut and 105 Ways of Preparing It for Human Consumption* (1917), National Agricultural Library, USDA.

162 Franz Eugen Köhler, *Arachis hypogaea* (1887), Köhler's Medizinal-Pflanzen.

168 *Big Buddha* by Nikki Lastreto.

169 *Sri Mukambika Temtple* by Chris Tucker.

169 *Ganesh at the Gate* by Nikki Lastreto.

180 Adapted from William Henry Fox Talbot, *Dandelion Seeds* (1858 or later), Photographs and Prints, Metropolitan Museum of Art.

181 Edward S Curtis, *Gathering Seeds--Coast Pomo* (ca. 1924), Photographs, Library of Congress.

182 Adapted from W. F. Kirby, *Hoary plantain, or lamb's tongue and perennial goosefoot*, British flowering plants, Page 41 (1906), Swallowtail Garden Seeds Collection, Flickr.

183 Adapted from John Stephenson and James Morss Churchill, *Scarlet pimpernel (Anagallis arvensis)* illustration from Medical Botany (1836), Rawpixel.

184 American Colony, *"Some seed fell among thorns"* (between 1934 and 1939), Photographs, Library of Congress.

189 Adapted from A. Groenendijk and Jacob Groenendijk, *Fire in the De Twee Leeuwen brewery in Rotterdam* (1782), Rijksmuseum.

190 Adapted from Jan Arendtsz, *After the Fire* (2014), Flickr.

195 Graphic by Tatiana Gómez Gaggero.

196 *Maule's seed catalogue* (1922), New York Botanical Garden Collection, BioDiversity Library.

206 Jean François Millet, *The Sower* (1851), Rijksmuseum.

207 Anonymous, *Rots in de wintertuin van de École Nationale Supérieure d'Horticulture in Versailles* (c. 1895 in or before c. 1900), Rijksmuseum.

208 *Vermoedelijk drie microscopische vergrotingen* (c. 1888 in or before 1893), Bulletin de la Société Impériale des naturalistes de Moscou (RP-F-2001-7-960), Rijksmuseum.

211 Isaac Chellman, *Recovered mountain yellow-legged frog*, National Park Service.

212 Illustrations by Addison Bowe.

217 Adapted from *Typic Plagganthrepts*, Soil Science, Flickr.

221 Adapted from Jan Punt, *Vignet met figuren bij een bijenkorf* (1732), Rijksmuseum.

222 *Trademark registration by Eckermann & Will for Bee Hive brand White Wax and Bleached Wax Goods* (1890), Photograph, Library of Congress.

223 Adapted from Ian Sane, *Community,* Flickr.

224,225 Thomas Halfmann, *Like Bee Hives* (2017), Flickr.

229 Adapted from E. L. Trouvelot, *Aurora Borealis: As observed March 1, 1872, at 9h. 25m. P.M.*, Rare Book Division, The New York Public Library.

232 *Bees Between pages 268 and 269*, from *The Home and School Reference Work, Volume I* (The Home and School Education Society, 1917), Scan by Sue Clark, Flickr.

233 Jack Delano, *Yabucoa, Puerto Rico. Sugar strikers picketing a sugar plantation* (1941), Photographs, Farm Security Administration, Library of Congress.

234 Jack Delano, *Cayey, Puerto Rico (vicinity). Cultivating tobacco at the PRRA (Puerto Rico Rehabilitation Administration) experimental area* (1941), Prints and Photographs, Library of Congress.

237 Jack Delano, *Untitled photo, possibly related to: Manati (vicinity), Puerto Rico. FSA (Farm Security Administration) borrowers and their families at a meeting held to discuss the distribution of land for tenant purchase farms* (1941), Prints and Photographs, Library of Congress.

238 Jack Delano, *Barranquitas (vicinity), Puerto Rico. Hands of an old woman working in a tobacco field* (1941), Prints and Photographs, Library of Congress.

242 *Press bulletin (Poston, Ariz.)*, September 20, 1942, Library of Congress, Serial and Government Publications Division.

245 Barr & Sugden, *Group of ornamental gourds from Barr & Sugden's spring seed catalogue and guide to the flower and kitchen garden* (1862), New York Botanical Garden, Biodiversity Heritage Library.

249 Dorothea Lange, *Migrant agricultural worker's family. Seven hungry children. Mother aged thirty-two. Father is native Californian. Nipomo, California. California Nipomo Nipomo* (1936), Photograph, Library of Congress.

250 Vera Bock, *Work pays America! Prosperity* (between 1936 and 1941), Prints and Photographs, Posters: WPA Posters, Library of Congress.

253 United Farm Workers, *UFW Boycott lettuce*, United States (1965–1980), Photographs, Library of Congress.

256 Upton Sinclair, "The Twelve Principles of Epic (End Poverty In California)" from *I, Governor of California and How I Ended Poverty* (1934), Mapping American Social Movements project at the University of Washington.

257 Adapted from Salvatore Trinchese, *Plate XV, Æolididae e famiglie affini del porto di Genova* (1877-1881), Museum of Comparative Zoology, Ernst Mayr Library, Harvard University.

260 *Don't Buy Bombs When You Buy Bread!* (between 1965 and 1975), Photograph, Yanker Poster Collection, Library of Congress.

261 *Outdoors amphitheater in Norris Park, close to recreation lodge see K 749. Split logs serve as benches, stage is paved with local flagstone, thickly planted shrubbery is used for wings and dressing rooms. It is used for informal theatricals, lectures, and meetings* (between 1933 and 1945), Farm Security Administration, Photograph,

Library of Congress.

262 Adapted from Anton Joseph von Prenner, *The Tower of Babel* (between 1683 and 1761), Prints, Metropolitan Museum of Art.

264,265 Adapted from Edward Hamilton, *Plate XVI* from *The flora homoeopathica: or, illustrations and descriptions of the medicinal plants used as homoeopathic remedies* (1852), Missouri Botanical Garden's Materia Medica, Biodiversity Heritage Library.

266 Original illustration by Katy Harris.

267-269 Illustration from *Brockhaus and Efron Encyclopedic Dictionary* (between 1890 and 1907), Wikipedia.

268 Adapted from F.D. Coburn, page 3 from *Alfalfa … : practical information on its production, qualities, worth, & uses, especially in the United States & Canada* (1912), Medical Heritage Library.

269 Adapted from François Lucas, *Ontwerp voor het inlegwerk van de kolf van een geweer* (ca. 1790–1800), Rijksmuseum.

273 *Lodge in Norris Park. This is one of the TVA (Tennessee Valley Authority) outdoor recreational resorts developed by the TVA in cooperation with the National Park Service and the CCC (Civilian Conservation Corps) of Norris Lake. The lodge contains facilities for dining and dancing and has a beautiful view of the lake from dining terrace. The park is developed in a rustic style native to the surrounding territory principally because the CCC program, which needed such opportunities, could furnish only low-skilled labor and practically no materials but those which could be procured by enrollees from local resources, such as timber and stone* (1933–1945), Farm Security Administration, Photograph, Library of Congress.

275 Kristen M. Caldon, *Grand Canyon National Park: Lookout Studio - Early Morning*, National Park Service.

276 Carol M. Highsmith, *Remains of the homes of ancient cliff dwellers in Canyon de Chelly (pronounced duh-SHAY), a vast park on Navajo tribal lands in northeastern Arizona, near the town of Chinle, that is now a U.S. national monument* (2018), Photographs, Library of Congress.

278 Vintage sweetest-scented hawthorn flower branch for decoration, Rawpixel.

281 *Youth Program (2019)*, Seattle Parks and Recreation, Seattle Parks.

282 Ray Filloon, *Class 3B and Class 3C. Keen Ponderosa Pine Tree Classification. Fremont National Forest, Oregon* (1939), Ray Filloon Collection, USDA Forest Service.

284,285 Photos by Briana Olson of work by Nelly Nguyen.

286 Adapted from Léon Becker, *Plate IX* from *Les arachnides de Belgique* (1882–96), Smithsonian Libraries, Biodiversity Heritage Library.

289 Dorothea Lange, *Dust storm near Mills, New Mexico* (1935), Farm Security Administration, Photographs, Library of Congress.

291 Elizabeth Twining, *Asteraceae. The Composite Tribe.* Illustrations from the *Natural Orders of Plants v.1*, (1868),

Swallowtail Garden Seeds Collection, Flickr.

292 Maria French, *Livre de Dessins De Joaillerie et Bijouterie* (1740–75), Books, Metropolitan Museum of Art.

294 *Attention boys and girls - $250.00 in prizes - miniature home building contest … May 3 1940* (1936–1940), Prints and Photographs, WPA Posters, Library of Congress.

295 Cassi Saari, *Tallgrass prairie flora Andropogon gerardii* (2013), Wikipedia.

296 Ernst Ferdinand Oehme, *Study of a Tree* (1832), Drawings and Prints, Metropolitan Museum of Art.

299 Adapted from Frederic Moore, *Plate 163* from *The Lepidoptera of Ceylon* (1880-1887), Harvard University, Museum of Comparative Zoology, Ernst Mayr Library.

301 Chris Hoving, *Smoke in the red pines* (2008), Flickr.

302 David Low, illustration from *On the Domesticated Animals of the British Islands: Comprehending the Natural and Economical History of Species and Varieties; the Description of the Properties of External Form; and Observations on the Principles and Practice of Breeding* (1845), the British Library.

305 Henri-Edmond Cross (Henri-Edmond Delacroix), *The Artist's Garden at Saint-Clair* (1904–1905), Drawings and Prints, Metropolitan Museum of Art.

307 George Washington Carver, *How to Build Up Worn Out Soils* (1905), National Agricultural Library, USDA.

309 Henry Gonzalez, *American Burying Beetle* (2019), US Fish and Wildlife Service.

310 Original work by Jo Zimny, Flickr.

315 *Arbuscular Mycorrhizal Fungi* by Roo Vandegrift.

316 Adapted from Georges Louis Leclerc Buffon, *Compléments de Buffon* (1838), Biodiversity Heritage Library, Smithsonian Libraries.

317 Charles Lippert and Jordan Engel, *Great Lakes from an Ojibwe perspective* (2015), The Decolonial Atlas.

318 Don Davis, *Cylinder Eclipse, Eclipse of the sun with view of clouds and vegetation*, NASA Ames Research Center.

319 *Hopi Corn Colored* by Michael Kotutwa Johnson.

320 Nick Townsend, *Ulva Isle, Scotland* (2011).

323 *Harper Lake, CA*, The Center For Land Use Interpretation, Morgan Cowles Archive Photo.

324 Adapted from *Spotprent op C. Seyn* (1817–1896), published by François Desterbecq, Rijksmuseum.

327 J.F. Castro, *The plague of locusts*, Wellcome Collection.

328 Alexander Shilling, *Schoffelende man in een weiland* (1923), Rijksmuseum.

330 Don Davis, *Torus Interior* (circa 1976), NASA Ames Research Center.

333 *Brick Moon Space Station Concept* (1869), NASA.

334 Photo No. A-33996, *Vertical Gun Range in horizontal loading position* (1965), NASA.

337 George Wharton James, *Interior of a Hopi Indian house at Oraibi, Arizona* (ca. 1900), California Historical Society Collection, University of Southern California Digital Library.

338 *Hopi boy planting corn* (1918), Special Collections, USDA National Agricultural Library.

339 Milton Snow, *Corn and melons. Sam Shingoitewas farm. 15 Miles S.W. Toreva Day School. Sam Shing [Shingoitewa] working* (1944), Digital Collection, University of Northern Arizona.

340,341 Maria Elena Peterson, *Many Hopi families teach their children how to raise corn by hand*, courtesy of Michael Kotutwa Johnson.

348 Adapted from Stu Smith, *Seaton park* (2017), Flickr.

352 Exhibition - drawings from Index of American Design Federal Art Project Works Progress Administration (1936–1938), Work Projects Administration Poster Collection, Library of Congress.

352,353 Seán Ó Domhnaill, *Farmland Near Ullapool. A cattle farm to the north of Ullapool* (2013), Flickr.

356 Wolcott and Christy, "The Joys of Farm Life," Larned Chronoscope, June 17, 1909.

357 William Morris, *Socialist League membership card* (1890), Digital Collections at the University of Maryland.

358 Catholic Worker, Vol. XXVIII No.3, October 1961, Courtesy of Jim Forest.

358 Dorothy Day, Courtesy of Jim Forest.

359 William Morris, *Fruit or Pomegranate* (1834–1896). Original from The Metropolitan Museum. Rawpixel.

360 Big Chipko Movement- Strongest Movement To Conserve Forests (2018), India Times, CC BY-SA 4.0.

360 Mark Garten, *An African chorus on climate change (Wangari Maathai)* (2009), United Nations.

362,363 Illustrations by Addison Bowe.

374 Vector illustrations by Peera, Rawpixel.

376 Pierre Joseph Hubert Cuypers, *Impressions of Leaves, Schetsboek met 23 bladen*, Rijksmuseum.

379 Pedro Szekely, *New Zealand countryside* (2019).

380 *Māori Farming, East Coast* (September, 1947), Archives New Zealand.

381 Gerrit Willem Dijsselhof, *Bare tree on a hillside property* (1876–1924), Rijksmuseum.

384,385 Gerard van Spaendonck and Louis Charles Ruotte, *Lavatère à grandes fleurs* from *Fleurs dessinées d'après nature* (1799–1801), Rijksmuseum.

386 *Bartholdi Park*, from the *Botanical Coloring Book*, United States Botanic Garden. This showcase for sustainable gardening practices includes a kitchen garden, a carnivorous plant bog, shrubs and trees with edible fruits, and many different native trees, grasses, and ferns.

387 *World Desert*, from the *Botanical Coloring Book*, US Botanic Garden. Some also grow long, sharp spines to keep thirsty animals away. Many desert plants have thick stems or fleshy leaves that store water.

399 Pierre Louis Dubourcq, *Kunstenaar aan het werk in de natuur* (in or before 1841), Rijksmuseum.

400 Original photo by Lydia Lapporte.

INDEX OF CONTRIBUTORS

June Alaniz
Writer, environmentalist
Edinburg, Texas

Emily Ryan Alford
Farmer, artist
Los Angeles, California
safeplaceforyouth.org

Natasha Balwit-Cheung
Writer, researcher, designer
Albuquerque, New Mexico
balwitcheung.com

Matsuo Bashō
Poet, essayist, traveler
Ishū, Japan and Mercury

Jo Ann Baumgartner
Executive director of Wild Farm Alliance
Watsonville, California
wildfarmalliance.org + bit.ly/BeneficialBirds

Catherine E. Bennett
Regenerative agriculturist
Among the Sumacs in DePeyster, New York
facebook.com/Milkweed-Tussock-
Tubers-933885123412173

Lauren August Betts
Beginning farmer, writer, eater
Sebastopol, California

Alice Blackwood
Ecologist, poet, performer
Bathurst, New South Wales, Australia

David Boyle
Author of *Back to the Land* and other books
Steyning in Sussex
david-boyle.co.uk

Addison Bowe
Farmer, artist
Rochester, Washington

Liz Brindley
Food illustrator
Santa Fe, New Mexico
printsandplants.com

Kaitlin Bryson
Artist, mycologist, activist
Los Angeles, California
kaitlinbryson.com

Adam Calo
Food systems researcher
Scotland
twitter.com/adamjcalo

Carlesto624
Visual artist
Jonesboro, Georgia
6hundred24.bigcartel.com

George Washington Carver
Regenerative farmer, inventor, educator
Tuskegee Institute

Brett Ciccotelli
Fisheries biologist
Downeast Maine
mainesalmonrivers.org

Cooking Sections
(Daniel Fernández Pascual and Alon Schwabe)
Spatial practitioners
London, UK
climavore.org

Ungelbah Davila
Writer, photographer
Bosque Farms, New Mexico
ungelbahdavila.com

Paola de la Calle
Interdisciplinary artist
San Francisco, California
paoladelacalle.com

Bernadette DiPietro
Visual Artist, writer, photographer
Ojai, California
bernadettedipietro.com

Charlotte Du Cann
Writer, editor, and co-director of
the Dark Mountain Project
Suffolk, UK
dark-mountain.net +
charletteducann.blogspot.co.uk

Nina Elder
Artist, researcher
Albuquerque, New Mexico
ninaelder.com

Jordan Engel
Mapmaker
Lenapehoking (Brooklyn, New York)

Iona Fox
Cartoonist
Chicago
instagram.com/iona_fox

Amy Franceschini
Artist, graphic designer
San Francisco, California
futurefarmers.com

Anamarija Frankić
Ecologist, oyster whisperer
Preko, Croatia
NORRI.ie

Zoë Fuller
Farmer, activist
Lazy Mountain, Alaska
singingnettle.farm

Kelly Garrett
Multidisciplinary artist
Brooklyn, New York
kellymgarrett.com

Frankie Gerraty
Biologist, storyteller
Santa Cruz, California
frankiegerraty.com

Melanie Gisler
Southwest director of Institute for Applied Ecology
Santa Fe, New Mexico
appliedeco.org

Christie Green
Mother, landscape architect, writer, hunter
Santa Fe, New Mexico
beradicle.com

Emily Haefner
Teacher, gardener
Dubois, Wyoming

Elizabeth Henderson
Organic farmer, writer
Rochester, New York
thepryingmantis.wordpress.com

Jon Henry
Farmer, artist
New Market, Virginia
thejonhenry.com

Alex Hiam
Author, illustrator
Putney, Vermont
alexhiam.com

Debbie Hillman
Soil-builder, food & farm strategist
Evanston, Illinois
foodfarmsdemocracy.net

Elizabeth Hoover
Professor, gardener
University of California, Berkeley
gardenwarriorsgoodseeds.com

Caroline Shenaz Hossein
Professor of business and society, founder of
Diverse Solidarity Economies Collective
Toronto, Canada
Caroline-Shenaz-Hossein.com +
@carolinehossein

Jamie Hunyor
Farmer-poet
Sierra Nevada foothills, California

William Hutchinson
NMDOT state landscape architect
Santa Fe, New Mexico
dot.state.nm.us/content/nmdot/en/Engineering_
Support.html#environment

Rob Jackson
Environmental scientist, poet
Stanford, California
jacksonlab.stanford.edu

Jenevieve
Artist, mother, healer
Atlanta, Georgia
jenevieve.info

Michael Kotutwa Johnson
Research associate,
Native American Agriculture Fund
Hopi Tutskwa

Dan Kelly
Poet, gardener
Tāmaki-Makaurau Auckland, Aotearoa
New Zealand
ateneoroa.home.blog

Rhamis Kent
Founder/director of Agroecological Natural
Technology Solutions
Cornwall, UK
antsregenerate.com

Hans Kern
Writer, illustrator, citizen's assembly advocate
Munich
swarming.global

Kuaʻāina Ulu ʻAuamo staff + contributing artists
Mark Lee (Holladay Photo), Scott Kanda, Kim Moa
Grassroots community-driven organization
Ka Pae ʻĀina o Hawaiʻi
kuahawaii.org

Lydia Lapporte
Student, weaver, educator
Bay Area, California
seaweedcommons.org

Nikki Lastreto
Writer, cannabis farmer
Laytonville, California
swamiselect.com

Amy Lindemuth
Landscape architect
Seattle, Washington

Charles Lippert
Ojibwe-speaker
Ne-zhingwaakokaag (Pine City, Minnesota)

Poppy Litchfield
Printmaker
Derbyshire, UK

Rob McCall
Journalist, fiddler
Brooklin, Maine

Michael McMillan
Ecological designer, educator
Mancos, Colorado
solsticesowndesigns.com

Alli Maloney
Editor, writer, photographer
Appalachia
allimaloney.com

Jessica Manly
Farmer at Hamburger Hill Farm, communications
director at National Young Farmers Coalition
Charlottesville, Virginia
youngfarmers.org

Austin Miles
Student
Ohio

Jacob Miller
PhD student
Manhattan, Kansas
jam199540.wixsite.com/personalsite

Nikki Mokrzycki
Artist, designer
Cleveland, Ohio
nikmoz.com

Jenni Monet
Investigative journalist, founder of *Indigenously*
Tribal citizen of the Pueblo of Laguna
jennimonet.com

Hollis Moore
Artist, landscape designer
Albuquerque, New Mexico
hollislmoore.com

Krisztina Mosdossy
Agricultural soil ecology PhD student
Montréal, Canada

Kirsten Mowrey
Writer, teacher, healer, guide
Ypsilanti, Michigan
kirstenmowrey.com

Maria Mullins
Southwest Seed Partnership coordinator
Santa Fe, New Mexico
Southwestseedpartnership.org

Yanna Mohan Muriel
Farmer-teacher
Utuado, Puerto Rico
allianceforagriculture.org

Nelly Nguyen
Photographer
Ho Chi Minh City, Vietnam
nellynguyen.com

Geoff Nosach
Artist, cultivator
Mercer, Maine

Lauren E. Oakes
Scientist, author, educator
Bozeman, Montana
leoakes.com

Tao Orion
Regenerative land stewardship consultant
Cottage Grove, Oregon
resiliencepermaculture.com

Alison Chow Palm
RN, homesteader
Midcoast Maine

Phoebe Paterson de Heer
Gardener, writer
South Australia

Colleen Perria
Savanna restorationist
Great Lakes region
opengrownschool.com

Lauris Phillips
Conservationist, artist
Humboldt County, California
laurisphillipsart.com

Nina Pick
Writer, oral historian, spiritual counselor
Egremont, Massachusetts
ninapick.com

Lily Piel
Photographer, digger
Belfast Ecovillage
lilypielphotography.com

Ben Prostine
Writer, agrarian
Soldiers Grove, Wisconsin

Molly Reeder
Botanical and food artist
Richmond, Virginia
mollyreeder.com

Quinn Riesenman
Trying to figure a few things out
Hesperus, Colorado
quintonriesenman@gmail.com

Rose Robinson
Future farmer, artist, poet, student
Newburyport, Massachusetts
@thelifeofmugs + thelifeofmugs.com

Ang Roell
Beekeeper, writer, facilitator
Great Falls, Massachusetts + Southern Florida
theykeepbees.com

Shamu Sadeh, Janna Siller, Rebecca Bloomfield
Adamah staff
Falls Village, Connecticut
adamah.org

Alejo Salcedo
Printmaker, tattooer
Atlanta, Georgia
alejosalcedo.com

M.T. Samuel
Poet
Alaska

Matthew Sanderson
Professor of sociology, anthropology, and social
work; editor of *Agriculture and Human Values*
Manhattan, Kansas
matthewsanderson.wordpress.com

Danielle Stevenson
Community scientist and toxicologist,
bioremediator, fungal emissary
Riverside, California
diyfungi.blog

The Submergence Collective
Artists, writers, stewards
New Mexico, Arizona, California
thesubmergencecollective.com

Makshya Tolbert
Writer, land worker, artist
Berkeley, California

Elena Vanasse Torres
Agrarian, land artist
Jayuya, Puerto Rico

Lisa Trocchia
Food systems scholar
Athens, Ohio
lisatrocchia.com

John Trudell
Poet, activist, musician
Earth
johntrudell.com

Tony N. VanWinkle
Faculty in sustainable agriculture & food systems
Sterling College
Craftsbury Common, Vermont

Emily Vogler
Teacher, designer
Rhode Island + New Mexico

Margaret Walker
Poet, novelist, scholar
African American Literary Hall of Fame

Karen Washington
Farmer, activist
Bronx, New York
karenthefarmer.com

Henry Weston
Director Of Forest Development,
Grants & Partnerships
Te Uru Rākau, Forestry New Zealand
teururakau.govt.nz

Elicia Whittlesey
Farmer, adjunct instructor
Mancos, Colorado
singthewinterwren.wordpress.com

Reid Whittlesey
Ecological restorationist
Santa Fe, New Mexico
riograndereturn.com

Carmela Wilkins
Social impact designer
Brooklyn, New York
carmelawilkins.com

Gerrard Winstanley
Agrarian communist
St. George's Hill, Surrey, England
diggers.org

Lucy Zwigard
Young farmer, cook, musician
Berkeley, California

SUBMISSIONS

At the end of every Almanac is . . . the start of the next Almanac!

Was your voice missing from Volume V? Have you discovered new strategies for repair, woken to post-Trumpian epiphanies, drafted a fresh, enduring manifesto for the rights of Earth?

Tune into our website, blog, newsletter, and social media to catch the call for submissions for Volume VI, "Adjustments and Accommodations." You can also share your visions for building, planting, seaweed farming, community land ownership, transformative finance, citizen science projects, rotational strategies, wildcrafting, rooftop gardens, seed migration, sleeping outside, and the art of the possible by email: almanac@greenhorns.org. To submit to our 2023 edition, reach out by March 2022.

We look forward to hearing from you.

ABOUT THE GREENHORNS

The Greenhorns works to promote, recruit, and support the next generation of farmers through grassroots media production. Our role is to explore the context in which new farmers face the world, through publications, films, media, and events—and by promoting the important work being done by so many organizations, alliances, trusts, and individuals around the world.

Starting in 2020, Greenhorns has adapted selected workshops into a digital production, part podcast and part magazine. The first series of EARTHLIFE features maritime and agricultural resources in Downeast Maine, highlighting the people who tend, conserve, protect, and adapt to this region. EARTHLIFE is a journey where we meet and speak with people doing the work and interpreting the potential of their landscapes.

Greenhorns is based in Downeast Maine along the Pennamaquan River in the old Pembroke Ironworks. Our campus is spread out around town with a carpentry shop, boat shop, mycological lab, agrarian library, and many living and art spaces. There's always something new getting going, and we welcome potential collaborators to come for a visit.

Stop by greenhorns.org to watch EARTHLIFE, download a guidebook, register for a seaweed webinar, or order all five volumes of the *New Farmer's Almanac*.

Join our mailing list for monthly news of naturalist trainings, EARTHLIFE releases, the next *New Farmer's Almanac*, and invitations to adventures on land, sea, and internet.